ISNM 110:
International Series of Numerical Mathematics
Internationale Schriftenreihe zur Numerischen Mathematik
Série Internationale d'Analyse Numérique
Vol. 110

Edited by
K.-H. Hoffmann, München; H. D. Mittelmann, Tempe;
J. Todd, Pasadena

Birkhäuser Verlag
Basel · Boston · Berlin

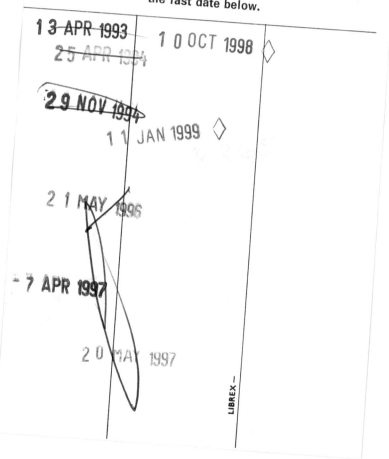

Software Systems for Structural Optimization

Edited by

H. R. E. M. Hörnlein
K. Schittkowski

1993

Birkhäuser Verlag
Basel · Boston · Berlin

Editors

Dipl.-Math. H. R. E. M. Hörnlein
Deutsche Aerospace
Unternehmensbereich Flugzeuge
Postfach 80 11 60
D-W-München 80
Germany

Prof. Dr. K. Schittkowski
Mathematisches Institut
Universität Bayreuth
Postfach 3008
D-W-8580 Bayreuth
Germany

A CIP catalogue record for this book is available from the Library of Congress, Washington D.C., USA

Deutsche Bibliothek Cataloging-in-Publication Data

Software systems for structural optimization / ed by H. R. E. M. Hörnlein; K. Schittkowski. – Basel ; Boston ; Berlin :
Birkhäuser, 1993
  (International series of numerical mathematics ; Vol. 110)
  ISBN 3-7643-2836-3 (Basel ...)
  ISBN 0-8176-2836-3 (Boston)
NE: Hörnlein, Herbert R. E. M. [Hrsg.]; GT

This work is subject to copyright. All rights are reserved, whether the whole or part of the material is concerned, specifically those of translation, reprinting, re-use of illustrations, broadcasting, reproduction by photocopying machine or similar means, and storage in data banks. Under § 54 of the German Copyright Law where copies are made for other than private use a fee is payable to »Verwertungsgesellschaft Wort«, Munich.

© 1993 Birkhäuser Verlag Basel
Printed from the authors' camera-ready manuscripts on acid-free paper in Germany
ISBN 3-7643-2836-3
ISBN 0-8176-2836-3

Contents

Preface VII

Structural optimization – a survey
H.R.E.M. Hörnlein 1

Mathematical optimization: an introduction
K. Schittkowski 33

Design optimization with the finite element program ANSYS[R]
G. Müller, P. Tiefenthaler and M. Imgrund 43

B&B: a FE-program for cost minimization in concrete design
W. Booz and G. Thierauf 57

The CAOS system
J. Rasmussen, E. Lund, T. Birker and N. Olhoff 75

Shape optimization with program CARAT
K.-U. Bletzinger, R. Reitinger, St. Kimmich and E. Ramm 97

DYNOPT: a program system for structural optimization weight minimum design
with respect to various constraints
D.W. Mathias and H. Röhrle 125

MBB-Lagrange: a computer aided structural design system
R. Zotemantel 143

The OASIS-ALADDIN structural optimization system
B. Esping, D. Holm and O. Romell 159

The structural optimization system OPTSYS
T. Bråmå 187

SAPOP: an optimization procedure for multicriteria structural design
H.A. Eschenauer, J. Geilen and H.J. Wahl 207

SHAPE: a structural shape optimization program
E. Atrek 229

STARS: mathematical foundations
P. Bartholomew and S. Vinson 251

Preface

Herbert Hörnlein, Klaus Schittkowski

The finite element method (FEM) has been used successfully for many years to simulate and analyse mechanical structural problems. The results are accepted or rejected by means of comparison of state variables (stresses, displacements, natural frequencies etc.) and user requirements. In further analyses the design variables will be updated until the user specifications are met and the design is feasible. This is the primary aim of the design process.

On this set of feasible designs, the additional requirement given by an objective function (e.g. weight, stiffness, efficiency, etc.) defines the structural optimization problem.

In recent years more and more finite element based analysis systems were extended and offer now optimization modules. They proceed from the design model as defined for structural analysis, to perform an internal adaption of design parameters based on formal mathematical methods. Despite of many common features, there are significant differences in the selected optimization strategy, the current implementation and the numerical results.

The basic intention of this book is to collect information on some well-known structural design optimization systems using an FEM analysis. The idea was born when organizing three COMETT-Seminars on Computer Aided Optimal Design in Thurnau (close to Bayreuth) between 1988 and 1992. To give the participants an impression on the performance of optimization software, part of each seminar was organized in form of a workshop, where developers of commercial and academic software presented their systems in form of lectures and computer demonstrations. They were invited to contribute a paper among others, so that - as we believe - we succeeded in obtaining a representative group of structural optimization systems.

An introduction into the basic mathematical model, i.e. the structural optimization problem, is presented in the first chapter. Also the importance of model formulation is motivated and outlined. All numerical optimization algorithms implemented in the software systems described, proceed from the same underlying optimization model, i.e. to minimize a nonlinear function subject to nonlinear inequality constraints, where all functions are differentiable. Thus we present also a very brief introduction into the mathematical optimization theory in the second chapter. Particularly the optimality criteria and the three mostly used optimization strategies are presented.

The table gives a brief review on the software systems in alphabetic order, which are considered in subsequent chapters:

Name	Organization
ANSYS	Swanson Analysis Systems Inc., Houston, USA
B & B	Fachbereich Baumechanik, Universität Essen, Germany
CAOS	Institute of Mechanical Engineering, Aalborg University, Denmark
CARAT	Institut für Baustatik, Universität Stuttgart, Germany
DYNOPT	Dornier Luftfahrt GmbH, Friedrichshafen, Germany
LAGRANGE	MBB-Flugzeuge, Ottobrunn, Germany
OASIS-ALADDIN	Alfgam Optimering AB, Stockholm, Sweden
OPTSYS	SAAB-SCANIA, Linköping, Sweden
SAPOP	Institut für Mechanik und Regelungstechnik, Universität Siegen, Germany
SHAPE	Engineering Mechanics Research Corp., Troy, USA
STARS	Defence Research Agency, Farnborough, England

The papers are organized in the following way:

- General outline and scope of the system
- Optimization model
- FEM techniques
- Sensitivity analysis
- Software organization
- CAE integration, pre- and postprocessing
- Applications and examples

The book is directed to engineers, mathematicians and scientific staff members who want to become familiar with modern procedures of structural optimization. The collected papers are mainly addressed to mechanical and civil engineering, but the underlying concepts of mathematical programming are useful in other disciplines as well. Basic knowledge of discretization methods (FEM) and matrix algebra is desirable.

The book also may be of interest for insiders and decision makers who want to be informed about new trends and the latest developments of software for their own needs.

Structural Optimization
- A Survey -

Herbert R. E. M. Hörnlein

Abstract. This article is addressed to beginners as well as to advanced students in the recently established discipline of *structural optimization*. Structural optimization is not a theory of its own, but it makes extensive use of theoretical results from several research disciplines. Mechanical engineering and mathematical programming theory is necessary to develop a programming system for structural optimization. Some mechanical engineering background is essential to realize the formulation of the design problem. To understand how to set up the design model the user will be guided carefully from a simple example to the important class of *displacement related constraints*. Furthermore a brief discription of more general design problems is given to introduce the scope of mechanical fields that can be managed by the tools of structural optimization. The consideration of the common mathematical formulation of all kinds of problems leads to a simultaneous solution process. Both the classical Optimality Criteria (OC) approach as well as the use of Mathematical Programming (MP) algorithms can be seen as an attempt to solve the *dual* problem formulation. Not only the age-old polemical dispute between the OC and MP school of thought can be saddled by understanding the *duality in structural design*, but also the Sequential Convex Programming (SCP) technique can be deduced from it. Finally, ideas are presented how to exploit some special mathematical properties in the design formulation.

1 Introduction

Over the last decades a growing interaction took place between analysis and synthesis of structures. The Finite Element Method (FEM) is an established simulation tool since fast computers have been available. Today almost all kinds of problems in structural mechanics can be evaluated by the many FEM–based codes which are available on the market.

Computer Aided Engineering (CAE) is the magic phrase in the community of engineers who have to simulate, analyze and verify the behaviour of specified structures. The main concept of CAE is to link the mighty capabilities of special purpose tools:

- Computer Aided Design (CAD) with incorporated, adaptive meshing covers the preprocess.

- The central task is the realistic, accurate simulation of the structures behaviour by FEM analysis codes.

- Eventually, the structural response will be prepared by the postprocessors i.e. plots, lists, tables and figures are used to interpret the deformations, stresses, modes, temperatures etc.

The entire process is commonly called *design process*, but unfortunately it neither designs the structure nor is it of any help to meet the users requirements. Usually the specification of structural problems will be done by *constants* and sometimes by quantities that are considered as *parameters*. The variation of these *free but fixed* parameters can be seen as the *design space*, but this is a remedy and it is a hopeless situation if more than a handful of parameters are taken into account and therefore too many analyses would have to be done.

What really is needed, is the introduction of a new entity, termed *variable*. With the help of design variables it is possible to convert the structural analysis formulation into a synthesis task which is called *structural optimization*. Instead of obtaining the structural behaviour for specified parameters, now the structural parameters are required for specified behaviour. Do you see the difference? It is the inverted problem. Therefore, in the past the structural optimization problem has often been termed as:

The reverse problem of structural mechanics

But optimization can effect even more than this. Beside a feasible design that satisfies the imposed constraints an additional goal can be established. The user of optimization tools is free to define an objective function that shall be minimized or maximized. This feature is new and the approach is completely different from the ridiculous attempt of enumeration mentioned above.

The first applications in structural optimization were tried by transforming the intuitive physical criteria of optimality into algorithms. This process led to nonlinear relations which had to be truncated in order to get easy to use redesign formulas. This approach is known as the Optimality Criteria method (OC). These methods are appropriate to provide an *optimal design* of very special structural optimization problems with sever restrictive assumptions. The *stress rationing* procedure e.g. leads to the famous *fully stressed design* (FSD), but this is optimal only if the structure is not statically redundant, if the used material has the same properties for all elements and if only stress constraints are imposed. Not even displacement constraints, which are in linear relationschip to the stress constraints can be managed simultaneously by this approach.

Today the *recurrence relations* (redesign formulas) are deduced from the Kuhn-Tucker-conditions [Ber/Ven 84]. Although these conditions are mathematically strict, the application on physically distinct constraints results in different recurrence relations which can be handled only within an *interactive* procedure. This approach remains doubtful as long as it is not known which of the constraints are active at the optimum.

The *simultaneous* consideration of physically different constraints was not possible until the Mathematical Programming methods (MP) were employed. The MP methods, which are independent of physics, supply the necessary generalization for the treatment of physically different restrictions.

Many algorithms of MP enables the engineer to attack his demanding challenge of structural optimization. The use of mathematical programming algorithms is becoming increasingly important. Linear and quadratic programming problems can be solved in a finite number of steps. However, the task of structural optimization is in general nonlinear and the suitable mathematical programming algorithms are iterative. Hence, there is a profound interest in the convergence analysis of these algorithms. Unfortunately convergence has only been proved under various and stringent conditions, such as for quadratic or at best for convex problems. For example, the *n–step–quadratic convergence* could be proved for the widely used Quasi–Newton method at least for asymptotically exact line search. This property indicates that the convergence speed is dependent on the number of variables, and this is precisely what can be observed by practical applications to large–scale design problems.

For a long time, attempts have been made to attack the problem inherent difficulties by remodelling the design task. Hence, the use of *reciprocal variables, suitable constraint approximations, linking of variables* and even the introduction of *cutting planes* and *move limits* is more or less the desperate attempt to make the *real* problem more attractive to the established MP methods. Why not the other way round?

Thanks to many years of experience in using MP methods, the experts in structural design have now such an intimate knowledge of MP theory that they ought to be self–confident enough to demand problem specific procedures from the MP experts. On the other hand, the school of MP should accept this challenge and signal its readiness for interdisciplinary cooperation.

Many MP codes have been developed during the seventies and eighties and a multitude of tests are available to assess these algorithms see e.g. [Schittk 80]. Here, efficiency, accuracy, reliability, ease of use and the dependency to the initial designs has been used as a ranking criteria. During the development of the in-house programming system *MBB–LAGRANGE* it has been realized that transparency, modularity and portability of the codes is at least from the same importance as the bench marks mentioned above. Since the test examples have very general properties, the conclusions of these test studies may not be applied immediately to structural optimization problems. Special studies for structural design problems are rarely carried out [Bel/Aro 84], [Bel/Aro 85] or [San/Rag 80].

To be prepared for the consultation with MP experts, the engineers should study the mathematical characteristics of their sort of problems. If e.g. the considered problem is known to be convex, many conclusions can be deduced about the solution as well as about the algorithm to be applied. Some important special cases of displacement related as well as fundamental frequency constraints with structural weight as an objective function have been shown to be convex by [Svanb 84].

Eventually, this survey article is supposed to encourage the design engineer to investigate his problem for special mathematical structure like *separability, convexity, homogeneity* and *bilinearity*. Some hints are given why these properties are of interest and how to exploit them. Concentration on constraints that can be expressed by linear combinations of displacements is done on purpose, because the main concern in standard design problems is addressed to stress, strain and explicit displacements. Even for special aircraft applications like stationary aeroelastic efficiency, twist and camber of wings are linear combinations of displacements.

2 Basic example

A two-bar truss as shown in fig. 1 will be used to demonstrate the elementary minimum weight/volume problem subject to displacement constraints. Only basic knowledge of FEM technique is required to realize this simple example.

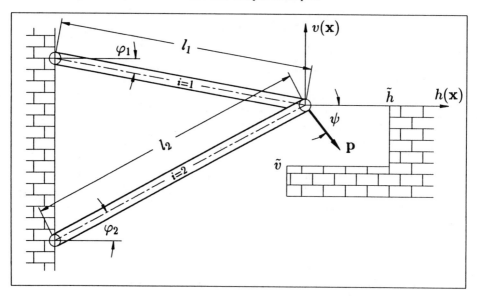

Figure 1: Two-bar truss with displacement constraints

The load vector $\mathbf{p}$ is defined as $\mathbf{p} = \left\{ \begin{array}{l} p_h = p \cdot \cos \psi \\ p_v = p \cdot \sin \psi \end{array} \right\}$ and the geometrical definitions

$-\frac{\pi}{2} \leq \psi \leq \varphi_1 \leq 0 < \varphi_2 < \frac{\pi}{2}$ imply the useful inequalities: $\left\{\begin{array}{c} 0 \leq \varphi_1 - \psi \leq \frac{\pi}{2} \\ 0 < \varphi_2 - \psi < \pi \\ -\pi < \varphi_1 - \varphi_2 < 0 \end{array}\right\}$

The mathematical properties of displacement functions realized by this example will be noted and in section 3 extended to more general large scale FEM structures. With the *element stiffness matrices*:

$$\mathbf{k}_i(x_i) = x_i \begin{bmatrix} c_i^2 & s_i c_i \\ s_i c_i & s_i^2 \end{bmatrix} \qquad i = 1, 2$$

where $c_i \stackrel{\text{def}}{=} \cos \varphi_i$; $s_i \stackrel{\text{def}}{=} \sin \varphi_i$ and for simplicity the *design variable*[1] $x_i \stackrel{\text{def}}{=} E \cdot A_i / l_i$ the assembled *global stiffness matrix*:

$$\mathbf{K}(\mathbf{x}) = \mathbf{k}_1(x_1) + \mathbf{k}_2(x_2)$$

and its invers

$$\mathbf{K}^{-1}(\mathbf{x}) = \frac{1}{|\mathbf{K}|} \begin{bmatrix} x_1 s_1^2 + x_2 s_2^2 & -x_1 s_1 c_1 - x_2 s_2 c_2 \\ -x_1 s_1 c_1 - x_2 s_2 c_2 & x_1 c_1^2 + x_2 c_2^2 \end{bmatrix}$$

with the determinant $|\mathbf{K}| = x_1 x_2 (s_1 c_2 - s_2 c_1)^2 = x_1 x_2 \sin^2(\varphi_1 - \varphi_2)$, the elastic equilibrium at the loaded hinge node yields:

$$\mathbf{K} \left\{ \begin{array}{c} h \\ v \end{array} \right\} = \left\{ \begin{array}{c} p_h \\ p_v \end{array} \right\} \quad \Rightarrow \quad \left\{ \begin{array}{c} h \\ v \end{array} \right\} = \mathbf{K}^{-1} \left\{ \begin{array}{c} p_h \\ p_v \end{array} \right\}$$

the explicit functions for horizontal and vertical displacements:

$$h(\mathbf{x}) = \frac{p}{x_1 x_2 \sin^2(\varphi_1 - \varphi_2)} \{\cos \psi (x_1 s_1^2 + x_2 s_2^2) - \sin \psi (x_1 s_1 c_1 + x_2 s_2 c_2)\} \qquad (1)$$

$$v(\mathbf{x}) = \frac{p}{x_1 x_2 \sin^2(\varphi_1 - \varphi_2)} \{\sin \psi (x_1 c_1^2 + x_2 c_2^2) - \cos \psi (x_1 s_1 c_1 + x_2 s_2 c_2)\}. \qquad (2)$$

By imposition of the *displacement constraints*:

$$0 < h(\mathbf{x}) \leq \tilde{h} \quad \text{and} \quad 0 > v(\mathbf{x}) \geq \tilde{v}$$

with the abbreviations:

$$F_h \stackrel{\text{def}}{=} \frac{p}{\tilde{h} \sin^2(\varphi_1 - \varphi_2)} > 0 \quad \text{and} \quad F_v \stackrel{\text{def}}{=} \frac{p}{\tilde{v} \sin^2(\varphi_1 - \varphi_2)} < 0$$

---

[1] Note, that only cross sectional areas A_1 and A_2 are design variables, all geometric entities are considered as parameters.

we get for the normalized horizontal displacement $\tilde{h}(\mathbf{x})$:

$$\tilde{h}(\mathbf{x}) = F_h \cos\psi \left(\frac{s_1^2}{x_2} + \frac{s_2^2}{x_1}\right) - F_h \sin\psi \left(\frac{s_1 c_1}{x_2} + \frac{s_2 c_2}{x_1}\right) \leq 1,$$

which is

$$\tilde{h}(\mathbf{x}) = \frac{1}{x_1} F_h s_2 \underbrace{\underbrace{(\cos\psi \cdot s_2 - \sin\psi \cdot c_2)}_{=\sin(\varphi_2 - \psi)}}_{=a_h > 0} - \frac{1}{x_2} F_h s_1 \underbrace{\underbrace{(\sin\psi \cdot c_1 - \cos\psi \cdot s_1)}_{=-\sin(\varphi_1 - \psi)}}_{=b_h \geq 0} \leq 1$$

or simplified

$$\tilde{h}(\mathbf{x}) = \frac{a_h}{x_1} - \frac{b_h}{x_2} \leq 1 \quad \Rightarrow \quad \frac{b_h}{x_2} \geq \frac{a_h - x_1}{x_1}$$

and the *feasible domain* in the 2-dimensional *design space* is explicitly bounded by:

$$\begin{aligned} x_2 &\leq \frac{b_h x_1}{a_h - x_1} = \frac{a_h b_h}{a_h - x_1} - b_h & \text{for} \quad x_1 < a_h \\ x_2 &\geq \frac{a_h b_h}{a_h - x_1} - b_h & \text{for} \quad x_1 > a_h. \end{aligned} \quad (3)$$

In the same way we derive the boundary for the vertical displacement $\tilde{v}(\mathbf{x})$:

$$\begin{aligned} x_2 &\geq \frac{b_v x_1}{x_1 - a_v} = \frac{a_v b_v}{x_1 - a_v} + b_v & \text{for} \quad x_1 > a_v \\ x_2 &\leq \frac{a_v b_v}{x_1 - a_v} + b_v & \text{for} \quad x_1 < a_v. \end{aligned} \quad (4)$$

The following conclusions from this example can be drawn:

- both displacements $h(\mathbf{x})$ and $v(\mathbf{x})$ are *linear in the reciprocal variables*
- both relations are *separabel*
- both relations are *homogeneous functions of the same degree*
- $v(\mathbf{x})$ is a *convex* function but $h(\mathbf{x})$ is a *non convex* function

and this implies a *non convex* feasible domain, see fig. 2.

By inspection of the Hessian matrix $\boldsymbol{D}_x^2 \tilde{h} \stackrel{\text{def}}{=} \left(\frac{d^2 \tilde{h}}{dx^2}\right)$, the horizontal displacement $\tilde{h}(\mathbf{x}) = a_h/x_1 - b_h/x_2$ can be shown not to be convex.

$$\boldsymbol{D}_x^2 \tilde{h} = 2 \begin{bmatrix} \frac{a_h}{x_1^3} & 0 \\ 0 & -\frac{b_h}{x_2^3} \end{bmatrix} \quad \Rightarrow \quad \boldsymbol{D}_x^2 \tilde{h} \text{ is } not \text{ positive semidefinite}$$

because of $a_h, b_h \geq 0$ and a twice continously differentiable function is convex if and only if its Hessian matrix is positive semidefinite.

Now we are looking for an explicit first order approximation which is *convex and separabel*. The benefit of separability will be discussed later. Introducing appropriate intervening variables $y_i \stackrel{\text{def}}{=} 1/x_i$, for all coordinates with negative derivatives $i \in I_- \stackrel{\text{def}}{=} \{i | \partial h/\partial x_i < 0\}$, yields:

$$\frac{\partial x_i}{\partial y_i} = -\frac{1}{y_i^2} \Rightarrow \frac{\partial h}{\partial y_i} = \frac{\partial h}{\partial x_i} \cdot \frac{\partial x_i}{\partial y_i} = -\frac{\partial h}{\partial x_i} \cdot \frac{1}{y_i^2}.$$

With linear taylor series expansion at $\tilde{\mathbf{x}}$

$$h(\mathbf{x}) = h(\tilde{\mathbf{x}}) + \sum_{I_+} \frac{\partial \tilde{h}}{\partial x_i}(x_i - \tilde{x}_i) + \sum_{I_-} \frac{\partial \tilde{h}}{\partial y_i}(y_i - \tilde{y}_i) + \mathcal{O}(|\Delta \mathbf{x}|^2)$$

where $I_+ \stackrel{\text{def}}{=} \{i | \partial h/\partial x_i > 0\}$ and after back substitution of $y_i = 1/x_i$ we get

$$\hat{h}(\mathbf{x}; \tilde{\mathbf{x}}) = h(\tilde{\mathbf{x}}) - \sum_i \left|\frac{\partial \tilde{h}}{\partial x_i}\right| \cdot \tilde{x}_i + \sum_{I_+} \frac{\partial \tilde{h}}{\partial x_i} \cdot x_i - \sum_{I_-} \frac{\partial \tilde{h}}{\partial x_i} \cdot \frac{\tilde{x}_i^2}{x_i} \quad (5)$$

as the desired explicit approximation with the demanded properties, which is well explained by [Fleury 86]. This approximation type is known as CONLIN.

To continue the example at the expansion point at $\tilde{\mathbf{x}} = (3,3)$ by setting the arbitrary parameters $a_h = 10, b_h = 4, a_v = b_v = 2$ we get

$$\tilde{h}(\mathbf{x}) = \frac{10}{x_1} - \frac{4}{x_2} \leq 1 \Rightarrow x_2 \leq \frac{40}{10-x_1} - 4 \quad \text{for} \quad x_1 < 10$$
$$\tilde{v}(\mathbf{x}) = \frac{2}{x_1} + \frac{2}{x_2} \leq 1 \Rightarrow x_2 \geq \frac{4}{x_1-2} + 2 \quad \text{for} \quad x_1 > 2$$

and for the convex approximation (5) at the same expansion point

$$\hat{h}(\mathbf{x}; \tilde{\mathbf{x}}) = \frac{10}{x_1} + \frac{4}{9}x_2 - \frac{8}{3} \leq 1 \Rightarrow \hat{x}_2 \leq \frac{33}{4} - \frac{90}{4} \cdot \frac{1}{\hat{x}_1}.$$

For comparison consider now the linearization $\bar{h}$ of h in fig. 2

$$\bar{h}(\mathbf{x}; \tilde{\mathbf{x}}) = 4 - \frac{10}{9}x_1 + \frac{4}{9}x_2 \leq 1 \Rightarrow \bar{x}_2 \leq \frac{10}{4}x_1 - \frac{27}{4}.$$

In fig. 2, $\mathbf{x}^*$ specifies the optimum of the *real* problem, while $\hat{\mathbf{x}}^*$ marks the optimum of the convex approximated problem. When the optimum $\hat{\mathbf{x}}^*$ has been calculated, a

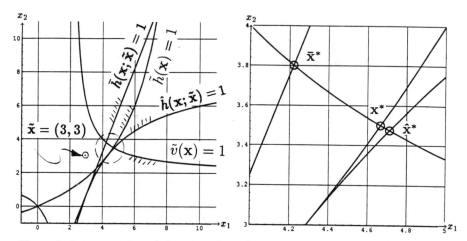

Figure 2: Interpretation of the example and comparison with the approximations

further approximation at $\hat{\mathbf{x}}^*$ for a new taylor expansion will be done. Continuation of these steps results in a Sequential Convex Programming (SCP). Unfortunately no convergence proofs for SCP exists or at least are not known to the author.

For the optimum $\mathbf{x}^*$ of this example, it is less remarkable that $\hat{\mathbf{x}}^*$ is a better approximation than $\bar{\mathbf{x}}^*$, but rather the *Conservativeness*: $\hat{h}(\mathbf{x}) > \bar{h}(\mathbf{x})$ for all $\mathbf{x} \in C$. This property implies immediately that the feasible domain of CONLIN approximations is a subset of the linearized feasible domain:

$$\widehat{M} \stackrel{\text{def}}{=} \{x \in C | \hat{h}(\mathbf{x}) \leq 1\} \subset \overline{M} \stackrel{\text{def}}{=} \{x \in C | \bar{h}(\mathbf{x}) \leq 1\}.$$

Moreover, it can be proved that within all *mixed approximations*

$$g_M(\mathbf{x}) \stackrel{\text{def}}{=} g(\tilde{\mathbf{x}}) + \sum_{I_1} \frac{\partial \tilde{g}}{\partial x_i}(x_i - \tilde{x}_i) - \sum_{I_2} \frac{\partial \tilde{g}}{\partial x_i} x_i^2 \left(\frac{1}{x_i} - \frac{1}{\tilde{x}_i}\right)$$

with $I_1 \cup I_2 = \{1, \ldots, n\}$ and $I_1 \cap I_2 = \emptyset$ CONLIN is the most conservative approximation.

3 Displacement related constraints

Now consider the usual minimum weight problem subject to constraints which can be expressed by linear combinations of displacements. We restrict ourselves to inequality constraints only, thus we can state the *primal problem*:

Structural optimization – a survey

$$\textbf{(P)} \quad \begin{aligned} \min f(\mathbf{x}) &= \mathbf{c}^T\mathbf{x}, \quad \mathbf{x} \in C \subset \mathbf{R}^n \\ \text{subject to} \quad g_j(\mathbf{x}) &= \mathbf{q}_j^T\mathbf{u}(\mathbf{x}) \leq \tilde{g}_j \quad \text{for} \quad j = 1,\ldots,m \\ \text{where e.g.} \quad C &= \{\mathbf{x} \in \mathbf{R}^n \,|\, x_i^l \leq x_i \leq x_i^u\} \end{aligned} \qquad (6)$$

$$\text{or equivalent} \quad \mathbf{x} \in \arg\{\min f(\mathbf{x}) \,|\, \mathbf{x} \in N\}$$
$$\text{where} \quad N \stackrel{\text{def}}{=} \{\mathbf{x} \in C \,|\, \mathbf{g}(\mathbf{u}(\mathbf{x})) \leq \tilde{\mathbf{g}}\}$$

$\mathbf{p}$ is the load vector and $\mathbf{u}$ is the vector of nodal displacements, implicitly obtained by the system equations of elastic equilibrium $\mathbf{K}\mathbf{u} = \mathbf{p}$.

With some additional restrictions to:

i) linear elements
$$\mathbf{k}_i(x_i) \stackrel{\text{def}}{=} x_i \cdot \mathbf{k}_i \Rightarrow \mathbf{K}(\alpha \cdot \mathbf{x}) = \sum_i \mathbf{k}_i(\alpha \cdot x_i) = \alpha \sum_i x_i \mathbf{k}_i = \alpha \cdot \mathbf{K}(\mathbf{x})$$

ii) linear related displacement constraints with no explicit dependency on the design variables
$$g(\mathbf{u}(\mathbf{x})) \stackrel{\text{def}}{=} \mathbf{q}^T\mathbf{u}(\mathbf{x}) \qquad \text{where } \mathbf{q} \text{ is a constant vector}$$

iii) invariant loads i.e. $\mathbf{p}$ is independent of $\mathbf{x}$: $\quad \mathbf{p}(\mathbf{x}) = \mathbf{p} =$ constant

and by introduction of reciprocal variables $y_i = 1/x_i$ we shall prove that the constraints ii) can be stated simply as

$$g(\mathbf{y}) = \mathbf{y}^T \nabla g. \qquad (7)$$

Proof: From i) we obtain for the reciprocal variables $\mathbf{y}$

$$\mathbf{K}(\mathbf{y}) = \sum_i \frac{1}{y_i} \mathbf{k}_i \quad \Rightarrow \quad \left(\frac{\partial \mathbf{K}}{\partial y_i}\right) = -\frac{1}{y_i^2} \mathbf{k}_i \qquad (*)$$

and from $\mathbf{K}\mathbf{K}^{-1} = \mathbf{I}$ with the product rule

$$\left(\frac{\partial \mathbf{K}^{-1}}{\partial y_i}\right) = -\mathbf{K}^{-1}\left(\frac{\partial \mathbf{K}}{\partial y_i}\right)\mathbf{K}^{-1} \stackrel{(*)}{=} \frac{1}{y_i^2}\mathbf{K}^{-1}\mathbf{k}_i\mathbf{K}^{-1}. \qquad (**)$$

Exploiting the system equations $\mathbf{u} = \mathbf{K}^{-1}\mathbf{p}$, the constraint ii) may be written with the constant vectors $\mathbf{q}$ and $\mathbf{p}$ as

$$g(\mathbf{y}) = \mathbf{q}^T \mathbf{K}^{-1} \mathbf{p}$$

and formal differentiation with respect to y_i yields

$$\frac{\partial g}{\partial y_i} \stackrel{(iii)}{=} \mathbf{q}^T \left(\frac{\partial \mathbf{K}^{-1}}{\partial y_i}\right) \mathbf{p} \stackrel{(**)}{=} \frac{1}{y_i^2} \mathbf{q}^T \mathbf{K}^{-1} \mathbf{k}_i \mathbf{K}^{-1} \mathbf{p} = \frac{1}{y_i^2} \mathbf{q}^T \mathbf{K}^{-1} \mathbf{k}_i \mathbf{u}. \qquad (***)$$

Hence, the constraint can be rewritten as

$$g(\mathbf{y}) = \mathbf{q}^T \underbrace{\mathbf{K}^{-1} \mathbf{K}}_{\mathbf{I}} \mathbf{u} = \mathbf{q}^T \mathbf{K}^{-1} \left(\sum_i \frac{1}{y_i} \mathbf{k}_i\right) \mathbf{u} = \sum_i y_i \cdot \frac{1}{y_i^2} \mathbf{q}^T \mathbf{K}^{-1} \mathbf{k}_i \mathbf{u}$$

$$\stackrel{(***)}{\Rightarrow} \quad g(\mathbf{y}) = \sum_i y_i \frac{\partial g}{\partial y_i} = \mathbf{y}^T \nabla g \qquad \square$$

Equation (7) seems to suggest that the displacement constraints are linear in the reciprocal variables **y** if the above mentioned conditions hold. This is of course not true since the gradient $\nabla g(\mathbf{y})$ is a function of the design variables. However, for statically determinate structures with independent internal forces **f** is the gradient a constant:

$$\mathbf{f}_p = \mathbf{k}_i(y_i)\mathbf{u}_p = \frac{1}{y_i}\mathbf{k}_i \mathbf{u}_p \quad \Rightarrow \quad \mathbf{u}_p = \mathbf{k}_i^{-1} \mathbf{f}_p y_i$$

$$\Rightarrow \quad \frac{\partial g}{\partial y_i} = \mathbf{q}_p^T \frac{\partial \mathbf{u}_p}{\partial y_i} = \mathbf{q}_p^T \mathbf{k}_i^{-1} \mathbf{f}_p = \text{const.}$$

where $(\cdot)_p$ is indicating the part of the structure which is controlled by the i-th design variable, all other components of the vectors **f, u, q** are removed to make the equation consistent. Finally, it should be mentioned that the use of reciprocal variables has also been successfully used for statically redundant structures. The most displacement related constraints become *'more linear'* if the reciprocal variable is applied, and it is worth trying Sequential Linear Programming (SLP) codes first.

4 Scope of 0ptimum design problems

4.1 Physical statement of problems

In the previous section only response based on linear relations to displacements has been considered. This covers the standard problems of *static response*:

- Displacements $\mathbf{u} = \mathbf{K}^{-1}\mathbf{p}$ $\Rightarrow$ $\mathbf{u}_l \leq \mathbf{u}(\mathbf{x}) \leq \mathbf{u}^u$
- Stresses $\boldsymbol{\sigma} = \mathbf{S}\mathbf{u}$ $\Rightarrow$ $\boldsymbol{\sigma}_l \leq \boldsymbol{\sigma}(\mathbf{x}) \leq \boldsymbol{\sigma}^u$
- Strains $\boldsymbol{\varepsilon} = \mathbf{B}\mathbf{u}$ $\Rightarrow$ $\boldsymbol{\varepsilon}_l \leq \boldsymbol{\varepsilon}(\mathbf{x}) \leq \boldsymbol{\varepsilon}^u$

where **S** and **B** are the constant stress and strain matrices and $(\cdot)_l$, $(\cdot)^u$ indicates the lower und upper restrictions.

Another important structural response is the local and global *buckling failure*:

- Buckling $(\mathbf{K} - \lambda \mathbf{K}_g)\mathbf{m} = 0 \Rightarrow \left\{ \begin{array}{ll} \text{static load} & \mathbf{p} \leq \text{buckling load} \quad \mathbf{p}(\lambda) \\ \text{static stress} & \boldsymbol{\sigma} \leq \text{buckling stress} \quad \boldsymbol{\sigma}(\lambda) \end{array} \right\}$

where $\mathbf{K}_g$ is the geometrical stiffeness matrix.

Furthermore the structure has to be checked for *dynamic failure*:

- Natural frequencies $\quad (\mathbf{K} - \omega^2 \mathbf{M})\mathbf{q} = 0 \quad \Rightarrow \quad \omega_l \leq \omega(\mathbf{x}) \leq \omega^u$

- Dynamic response $\quad (\mathbf{K}\mathbf{u} + \mathbf{C}\dot{\mathbf{u}} + \mathbf{M}\ddot{\mathbf{u}} = \mathbf{F}(t)) \quad \Rightarrow \quad \left\{ \begin{array}{l} \mathbf{u}_l \leq \mathbf{u} \leq \mathbf{u}^u \\ \dot{\mathbf{u}}_l \leq \dot{\mathbf{u}} \leq \dot{\mathbf{u}}^u \\ \ddot{\mathbf{u}}_l \leq \ddot{\mathbf{u}} \leq \ddot{\mathbf{u}}^u \end{array} \right\}$

where **M** is the mass matrix and **C** the damping matrix.

Finally, the interactions between the elastic forces the dynamic forces and the airloads are of concern to the aircraft engineers when reviewing *aeroelastic behaviour*:

- Flutter speed $\quad (\mathbf{K} - \omega^2 \mathbf{M} - \mu \mathbf{A})\mathbf{q} = 0 \quad \Rightarrow \quad v_f \leq v_f(\mu(\mathbf{x}))$
- Aeroelastic efficiencies $\quad \mathbf{K}\mathbf{u}_e = \mathbf{P} + \mathbf{A}\mathbf{u}_e \quad \Rightarrow \quad \varrho_l \leq \varrho(\mathbf{u}_e, \mathbf{u}_r) \leq \varrho^u$
- Aeroelastic divergence $\quad (\mathbf{K} - \mu \mathbf{A})\mathbf{q} = 0 \quad \Rightarrow \quad v_d \leq v_d(\mu(\mathbf{x}))$

where **A** is a transformed aerodynamic influence matrix, $\mathbf{u}_e$, $\mathbf{u}_r$ are the elastic and ridgid deflections and v_f, v_d are the flutter and divergence speed.

As shown above these state variables i.e. response of the structure, are implicitly dependent on the design variables **x**. The inequalities for the state variables have to be specified by the user and will be imposed as constraints on the design problem.

A definition of the objective function completes the optimum design problem. Beside the usual functional for weight or volume all kinds of state variables, or combinations, can be used as an objective function. Today the FEM simulation is spraed over all engineering disciplines and various structural responses may be used as constraints as well as for the objective function. Temperature, energy expressions, impact and acoustics are only some state variables which are of general interest in civil engineering.

4.2 Classification of problems

One possible classification of problems is to divide them according to the mathematical characteristics of the system equations. Some properties of the problem related functions are mentioned in section 8 and it is shown how to exploit their mathematical structure. For real life tasks it is rather difficult to observe a property which can be turned to an advantage.

A second more common sorting of the design problems is given by the *design space*. Figure 3 illustrates the potential range of the design space by a bridge design. Starting from the lowest level in this hierarchy, *sizing of elements* is the treatment of cross–sectional dimensions only. The structural system i.e. the idealization of an FE structure and its geometrical layout has to be established previously. This type of *sizing variables* is state of the art and can be handled by all existing programming systems for structural optimization.

The second level of design variables is called *shape variables*. In this case the structural system is fixed in advance but the geometrical entities as well as the sizing variables remain to be chosen optimally. Shape variables are coordinates of the nodes itself or parameters of *shape functions* that are related to the nodal positions.

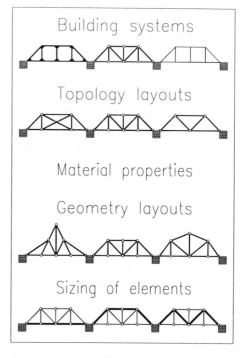

Figure 3: Classification of problems

Shape design has been treated extensively with FE structures for some years, see [Van/Mos 72], [Peder 73], [Esping 83], [Bot/Ben 84] and [Fleury 85]. The most difficult job within shape optimization is the adaptive mesh generation and refinement respectively.

Dealing with *material properties* as variables is possible by many programming systems. However, there are varying opinions as to which material parameters should be considered as variables, e.g. aeroelastic tayloring by means of composite material is still a challenging task. On the one hand, many laminates with fixed fibre orientations and variable thicknesses can be used to determine the suitable fibre direction. On the other hand, the fibre orientation angles *and* the thicknesses may be applied as design variables, but this can lead to serious nonlinearities and non convex or even not simply connected feasible domains.

The top level of optimization variables are the *topology variables*. Here, nothing of

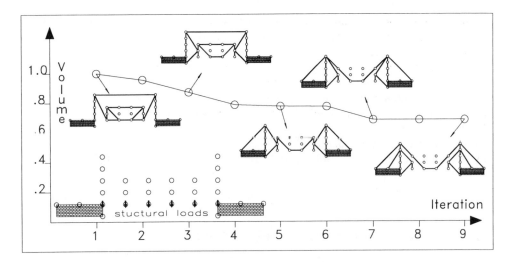

Figure 4: History for a topological bridge optimization

the structure is specified beforehand except the mission and the *building system*. The choice of the building system is more or less a qualification determined by tradition or by engineering genius. No decision criteria are known that can be used for automated designing. Coming back to the arrangement of elements. Topology has been a subject of research for more than 100 years. Among the most important studies are those of [Maxwe 1869] and [Miche 1904], whose ideas were first transformed into usable algorithms in 1964, see [Dorn et al. 64], and [Hemp 64]. Originally, all the work done to answer the question *'which nodes are connected by elements?'* was limited to skeletal structures on grid–like continua. Recent investigations in layout theory have been concerned with other types of elements. A new technique, called *homogenization method* is gaining ground since the end of the eighties, see [Bendsøe 89] [Ben/Kik 88], [Suz/Kik 91] and [Kik/Suz 91]. An exellent paper on topology optimization has been presented by [Bendsøe 92]. The importance of topology optimization is not yet widely realized. Current research activities should be observed to understand this largest potential improvement for optimum structural design.

A generic design of a bridge is shown in fig. 4 to give an impression of topology optimization. This result has been generated by the author in 1979 with an algorithm based on the early approach which is known as *ground structure* technique. On a grid selection of $g = 30$ nodes a total of $\binom{30}{2} = 435 = m$ rod elements are needed to connect all pairs of nodes. Starting from this global structure by removing any number of elements, a total of $2^{435} \approx 10^{130}$ combinations of rods are possible. Now, it can be proved that for single loaded truss structures with strength and buckling constraints, the minimum weight/volume problem results in a statically determinate structure. All member forces of non redundant structures can be calculated by equilibrium equations only. For the plane truss structure in fig. 4 we have $2 \cdot g = 60$ equations –two for each node– and therefore 60 rod elements –including support elements– are necessary to form a statically determinate structure. This observation was the basic idea for an algorithm which selects $2 \cdot g$ rod elements that forms a non redundant structure. Hence, among the 2^{435} rod combinations, at most $\binom{435}{60} \approx 10^{75}$ statically determinate structures are possible. But this number is still too large for enumeration of all structures. Reformulation of the topology design problem in terms of member forces resulted in a simple Linear Programming problem (LP) and the optimal topology could be found by only 9 iteration steps. Unfortunately this algorithm can neither be extended to multi loaded structures nor to displacement constraints, because these topologies are in general statically indeterminate.

5 Modelling and solving the problem

5.1 Set up the problem formulation

The formulation of the optimum design problem is at least from the same importance as the relevant choice of the optimization algorithm. Since the algorithms of MP have been taylored for special features of the functions, the problems should be adopted by clever remodelling. This is in general not straightforward or even impossible. A profound knowledge of the algorithms range of adaptability is obligatory. However, a lot of experience and understanding has been published on this topic.

Nevertheless, it is still more complex to understand the problems nature. Constraint functions which are defined implicitly can not be *seen* to be homogeneous, separable, bilinear or even convex. At best a function can be identified to be linear or nonlinear. No–one should be surprised whenever the algorithm fails to solve a new typ of problem. For almost all scientific applications, linearization is in permanent use, because everybody knows the trouble with *nonlinearities*. This is how to show Sequential Linear Programming (SLP) in its true light. Not only linearization is widely spread, but also nonlinear approximation as presented in section 7.

Appropriate reformulation is not always to achieve by approximation, but sometimes by remodelling. Recall the basic example of fig. 1 with different displacement constraints. Let $\mathbf{u}^T(\mathbf{x}) = (h(\mathbf{x}), v(\mathbf{x}))$ and $\tilde{\mathbf{u}} = (\tilde{u}, \tilde{v})$ then the scalarproduct $\mathbf{p}^T\mathbf{u}(\mathbf{x}) \leq \lambda \cdot \mathbf{p}^T\tilde{\mathbf{u}}$ combines the two constraints into one which can be seen as an approximate design model of the original basic example, see fig. 5 and compare with fig. 1.

The latter problem formulation seems to be even more reasonable. However, the real effect of this modification is the convexity of the resulting constraint function $g(\mathbf{x})$:

Structural optimization – a survey

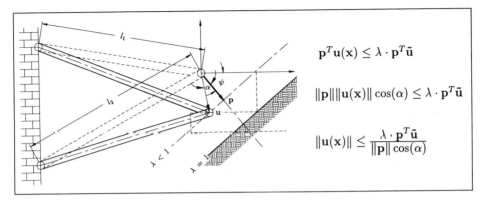

Figure 5: Model modification for the basic example

$$g(\mathbf{x}) \stackrel{\text{def}}{=} \mathbf{p}^T \mathbf{K}^{-1}(\mathbf{x}) \mathbf{p} \leq \lambda \cdot \mathbf{p}^T \tilde{\mathbf{u}} \stackrel{\text{def}}{=} \tilde{g}$$

Such constraints are called *symmetric displacement constraints*. The convexity of $g(\mathbf{x})$ is simply proved by showing the positive semidefiniteness of the Hessian matrix $\boldsymbol{D}_x^2 g = \left(\frac{\partial^2 g}{\partial x_i \partial x_j}\right)_{i,j=1,\ldots,n}$. Stiffness designs subject to these symmetric displacement constraints are easy to detect because it is a global minimum which should be found by all MP procedures. This is only one example of appropriate model modification. Small changes in problem formulations might result in favourable effects.

The general formulation of the structural optimization task can be stated as *nonlinear programming problem* (NLP):

$$
\begin{array}{lll}
\text{minimize} & f(\mathbf{x}) & \mathbf{x} \in \mathbf{R}^n & \text{(objective function)} \\
\text{subject to} & g_j(\mathbf{x}) \leq 0 & j \in J = \{1, \ldots, m\} & \text{(behavioural constraints)} \\
& \mathbf{x}_l \leq \mathbf{x} \leq \mathbf{x}_u & & \text{(side constraints)}
\end{array}
\quad (8)
$$

Numerous systems for structural optimization are available on the market, supplemented by many user options to assist the design engineer during the set up of his problems. Although the concepts and in particular the optimization codes are different, some of the common program features can be summarized as follows.

Variable linking provides the utalization of serial parts, structural symmetry and fabricational requirements. In mathematical terms, variable linking relates the many strutural variables say $\mathbf{t}$ and the user selected design variables $\mathbf{x}$ by affine linear transformation

$$\mathbf{t}(\mathbf{x}) = \mathbf{a} + \mathbf{A}\mathbf{x}$$

where $\mathbf{t}, \mathbf{a} \in \mathbf{R}^{\tilde{n}}, \mathbf{x} \in \mathbf{R}^n$ and $\mathbf{A} \in \mathbf{R}^{\tilde{n} \times n}$ is the adjacent matrix which identifies and linkes the structural members.

Intervening variables are used to change the characteristics of the functions. As already shown in section 3, the use of reciprocal variables is smoothing the usual strength and stiffness constraints.

Explicit approximation of constraints, reflects the basic idea of substituting the inaccessible, implicit relations locally by suitable explicit functions. Linear, hyperbolic as well as hybrid forms are in use for different types of constraints, see [Fleury 89a], [Fleury 89b], [Prasad 83] and [Schm/Far 74]. First order approximations which are convex and separable are utilized in *Dual concepts*, see also section 7.

Global approximation by means of reduced basic functions is also employed to represent structural response in the extended domain of the design space.

For large scale problems with many *local* constraints like stress and buckling, normally only a few *global* constraints, such as flutter speed or aeroelastic efficiencies are imposed. Decomposition of those problems have been tried since the early seventies. The splitting of large scale problems into small, nested problems is known as *multi-level* concept. This heuristic decomposition does not ensure that the true optimum will be found, but it makes large scale problems solvable and leads in general to near optimum designs. Noting that the success of multi-level approach depends greatly on the user's physical understanding. Some multiple constrained structures, such as wings, frames and ship designs have been solved successfully, see [Schm/Can 84], [Sob et al. 85] and [Hug et al. 80].

When the designer is challenged by more than one objective, he has to set all but one as constraints. This situation is often referred to as *trade-off* studies. *Vector optimization* allows the consideration of several objectives at the same time. Because it is not to expect that a design is optimal for all objectives a different statement known as *Pareto optimality* is utilized now. A design is called Pareto optimal if no simultaneous improvements for all objectives are possible. Hence, the designer must decide on which of the objectives he wans to give up in favour of the remainig objectives. In practice all methods that handle *multi-objective* problems are based on scalarization and the use of Pareto optimality criteria. In terms of decision theory, a scalar preference function replaces the decision maker, see chapter 10 in [Carmich 81], [Esch et al. 85] and [Stadler 84].

5.2 Solving the design problem

It is important to know by what decicions the design process is monitored. First of all, the user has to know when to stop the process. The most important termination condition is the stationarity of the Lagrangian function, which is called the *Kuhn–Tucker condition*. For details see the next chapter in this book: Mathematical Programming –An Introduction–. Understanding of *Duality* is helpful to realize when the optimum is reached, see also section 6.

Beside the mathematical conditions, which are implemented inside the MP codes, the user has to control the iterative process by checking the *relative alterations* of the objective function

$$\frac{|f(\dot{\mathbf{x}}^{\nu+1}) - f(\mathbf{x}^{\nu})|}{|f(\mathbf{x}^{\nu+1})|} \leq \varepsilon_f \qquad (9)$$

and the design variables

$$\frac{\|\mathbf{x}^{\nu+1} - \mathbf{x}^{\nu}\|}{\|\mathbf{x}^{\nu+1}\|} \leq \varepsilon_x \qquad (10)$$

as well as the *feasibility* of the design by checking the constraint vioalations[2]

$$\max_{j \in J} g_j(\mathbf{x}^{\nu}) \leq \varepsilon_g \qquad (11)$$

The problem dependent termination values $\varepsilon_f, \varepsilon_x$ and ε_g in (9) (10) and (11) have to be chosen carefully by the user. In addition the total number of iterations should be restricted to ensure termination if the design process is not converging.

Active set strategy divides the constraints J into *active*[3] JA and *passive* JP

$$JA \stackrel{\text{def}}{=} \{j \in J : g_j(\mathbf{x}) \geq -p_a\}$$
$$JP \stackrel{\text{def}}{=} \{j \in J : g_j(\mathbf{x}) < -p_a\}$$

where p_a can be seen as a percentage of violation if the constraints are normalized to get rid of the numerical effects of the physical units. A permanent change of active and passive constraints takes place throughout the design process. The *less active* constraints have less guidance on the concurrent design changes and may be therefore temporarily neglected. Suitable deletion of passive constraints accelerates the process, but the handling of constraint deletion is not simple. Small changes in the design can result in a completely altered active set. This may cause oscillation of the process.

During the design iterations the *Lagrangian multipliers* can be updated by checking the Lagrangian function for stationarity $\boldsymbol{D}_x L(\mathbf{x}*, \lambda*) = 0$ and the complementary slackness conditions $\lambda_j * g_j(\mathbf{x}*) = 0$ for $j \in J$. Close to the optimum, the estimated Lagrangian multipliers may be used as indicator for active constraints. Simple interpretation of the first order necessary Kuhn–Tucker conditions shows that active constraints are identified by positive multipliers.

However, the determination of active constraints is of main concern for all mathematical programmimg codes, but the information of which of the constraints are active or will become active is more or less directly used for design modifications. As already mentioned in the introduction, there are two schools of thought which take care of structural optimization.

---

[2] Note that the constraints may be defined as $g_j(\mathbf{x}) \geq 0$.
[3] In MP theory only $g_j(\mathbf{x}) = 0$ are considered as active.

The school of *Optimality Criteria* is using the iterative scheme

$$\begin{aligned} x_i^{\nu+1} &= \Phi(\mathbf{x}^\nu, \lambda^\nu) \quad i = 1, \ldots, n \\ \lambda^{\nu+1} &= \Psi(\mathbf{x}^\nu) \end{aligned}$$

where $\lambda^{\nu+1} \in \mathbf{R}^{|JA|}$ is an estimation of the Lagrangian multipliers. The iteration functions Ψ are deduced from the complementary slackness condition and the *recurrence relations* Φ are developed from the stationary conditions. Thus this approach needs to be under interactive control of the user to appoint the physically dependent iteration functions. Usually this is managed by varying *active* and *passive variables* x_i.

The physically independent more general school of *Mathematical Programming* follows the iterative scheme

$$x_i^{\nu+1} = \mathbf{x}^\nu + \alpha^\nu \mathbf{s}^\nu$$

where $\mathbf{s}^\nu \in \mathbf{R}^\nu$ is called *search direction* and $\alpha^\nu \in \mathbf{R}_+$ is a scalar *step size*. The crucial point with Mathematical Programming is the clever determination of the search direction **s**. Also here the knowledge of active constraints is utilized in order to speed up the process. The calculation of step size α is a simple matter, but it may require many stuctural analyses. Hence the step size procedure must be very efficient.

Finally, the *design sensitivity* analysis should be mentioned as an indispensable part of the overall design process. Not only the Mathematical Programming approach is using the gradients, but also for the evaluation of the recurrence relations in the Optimality Criteria scheme. The people of the OC school pretend not to make use of derivatives. But many terms of *strain energy density* are needed for almost all recurrence relations, and hence virtual displacements have to be calculated with the same effort as the derivatives of the constraint functions. Most of the computing time is spent in calculating the sensitivity analyses. Efficient and accurate calculation methods for gradient evaluations deserves particular attention. In resent years sensitivity analysis for structural response has been accepted as an individual discipline, see [Haf/Bar 91] and for an extensive reference list [Haf/Ade 89].

6 Duality in structural optimization

Considering again the *primal problem* (**P**) with inequality constraints only and *convex functions*

$$f : \mathbf{R}^n \longrightarrow \mathbf{R} \qquad \mathbf{g} : \mathbf{R}^n \longrightarrow \mathbf{R}^m$$

defined on the convex set[4] $C = \{\mathbf{x} \in \mathbf{R}^n | x_i^l \leq x_i \leq x_i^u\}$.

---

[4] C can be seen as the explicit restrictions on the design variables by means of min / max gauges with lower and upper bounds $x_i^l, x_i^u \geq 0$.

By introducing the *Lagrangian function* $L : \mathbf{R}_+^n \times \mathbf{R}_+^m \longrightarrow \mathbf{R}$

$$L(\mathbf{x}, \boldsymbol{\lambda}) \stackrel{\text{def}}{=} f(\mathbf{x}) + \boldsymbol{\lambda}^T \mathbf{g}(\mathbf{x}) \tag{12}$$

we define the *saddle points*:

A vector $(\mathbf{x}^*, \boldsymbol{\lambda}^*) \in \mathbf{R}_+^n \times \mathbf{R}_+^m$ is said to be a *saddle point* of L if

$$L(\mathbf{x}^*, \boldsymbol{\lambda}) \leq L(\mathbf{x}^*, \boldsymbol{\lambda}^*) \leq L(\mathbf{x}, \boldsymbol{\lambda}^*) \qquad \text{for all} \quad \mathbf{x}, \boldsymbol{\lambda} \succeq \mathbf{0}.$$

The classical *saddle point theorem*:

$$\text{If} \quad (\mathbf{x}^*, \boldsymbol{\lambda}^*) \quad \text{is a \emph{saddle point of} } L, \text{ then } \mathbf{x}^* \text{ solves } (\mathbf{P}) \tag{13}$$

gives a useful sufficient condition for the solution of (**P**). This statement even holds without convexity of f and $\mathbf{g}$ but it provides 'only' a sufficient condition. To make sure that each solution $\mathbf{x}^*$ of (**P**) corresponds to a saddle point of L, we need to know by what additional conditions the conversion of the saddle point theorem remains true.

Indeed, there are rules which are called constraint qualifications or regularity conditions[5] —ruling out certain pathological cases— for which holds the
Kuhn – Tucker – Theorem:

$$\begin{array}{c} \mathbf{x}^* \text{is the solution of the \emph{convex primal problem} } (\mathbf{P}) \\ \text{if and only if} \\ \text{a unique Lagrangian multiplier} \quad \boldsymbol{\lambda}^* \in \mathbf{R}_+^m \quad \text{exists} \\ \text{such that } (\mathbf{x}^*, \boldsymbol{\lambda}^*) \text{ is a saddle point of } L \end{array} \tag{14}$$

This result justifies the new concept to attack the problem (**P**) via the Lagrangian function. An alternative formulation of a saddle point is the solution of the *min–max problem*:

$$\min_{\mathbf{x} \geq 0} \max_{\boldsymbol{\lambda} \geq 0} L(\mathbf{x}, \boldsymbol{\lambda}) = \max_{\boldsymbol{\lambda} \geq 0} \min_{\mathbf{x} \geq 0} L(\mathbf{x}, \boldsymbol{\lambda}). \tag{15}$$

A general saddle surface is illustrated in fig. 6, but the Lagrangian saddle surface of our problem under consideration is of course convex in $\mathbf{x}$ for any $\boldsymbol{\lambda} \geq \mathbf{0}$ and linear in $\boldsymbol{\lambda}$ for any fixed $\mathbf{x}$, see fig. 7 for comparison.

The theorems (13) and (14) suggest the investigation of the Lagrangian function for saddle points and the *reformulation of the min–max problem* (15) due to [Stoer 63]:

$$\max_{\boldsymbol{\lambda} \geq 0} L(\mathbf{x}, \boldsymbol{\lambda}) \qquad \text{s.t.} \quad \mathbf{x} \in \arg \left\{ \min_{\mathbf{x} \in C} L(\mathbf{x}, \boldsymbol{\lambda}) \right\} \tag{16}$$

---

[5] One of these is the popular *Slater condition*: $\text{int}(N) \neq \emptyset \quad \Leftrightarrow \quad \exists \mathbf{x} \in C : \mathbf{g}(\mathbf{x}) < 0$

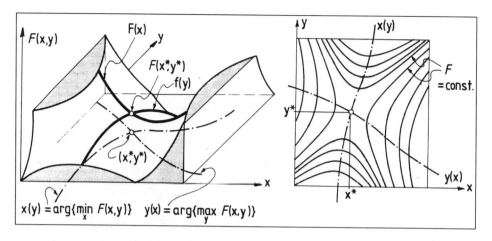

Figure 6: General saddle surface as min–max problem

This motivates the definition of the

valley : $l(\boldsymbol{\lambda}) \stackrel{\text{def}}{=} L(\mathbf{x}(\boldsymbol{\lambda}), \boldsymbol{\lambda})$ where $\mathbf{x}(\boldsymbol{\lambda}) = \arg \{\min L(\mathbf{x}, \boldsymbol{\lambda}) | \mathbf{x} \in C\}$

and *hill* : $\mathcal{L}(\mathbf{x}) \stackrel{\text{def}}{=} L(\mathbf{x}, \boldsymbol{\lambda}(\mathbf{x}))$ where $\boldsymbol{\lambda}(\mathbf{x}) = \arg \{\max L(\mathbf{x}, \boldsymbol{\lambda}) | \boldsymbol{\lambda} \geq \mathbf{0}\}$.

Thus we have immediately lower and upper bounds for any $\mathbf{x}$, $\boldsymbol{\lambda}$:

$$l(\boldsymbol{\lambda}) \leq L(\mathbf{x}, \boldsymbol{\lambda}) \leq \mathcal{L}(\mathbf{x}).$$

Now, the problem in question is: Which is the most efficient way to the saddle point, maximizing $l(\boldsymbol{\lambda})$ or minimizing $\mathcal{L}(\mathbf{x})$?

When considering the argument $\boldsymbol{\lambda}(\mathbf{x})$ of the *hill* function $\mathcal{L}(\mathbf{x})$ the linearity of $L(\mathbf{x},\boldsymbol{\lambda})$ in $\boldsymbol{\lambda}$ yields by inspection of (12)

$$g_j(\mathbf{x}) > 0 \quad \Rightarrow \quad \lambda_j(\mathbf{x}) = \infty$$
$$g_j(\mathbf{x}) < 0 \quad \Rightarrow \quad \lambda_j(\mathbf{x}) = 0.$$

Hence, $\lambda_j(\mathbf{x})$ is discontinuous on the boundary $g_j(\mathbf{x}) = 0$ of the feasible domain. Therefore the multiplier $\boldsymbol{\lambda}^*$ can not be determined by minimizing the upper bound function $\mathcal{L}(\mathbf{x})$. This approach can be shown to be the original *primal problem*:

$$\mathbf{x} \in \text{int}(N) \quad \Rightarrow \quad g(\mathbf{x}) < 0 \quad \Rightarrow \quad \boldsymbol{\lambda}(\mathbf{x}) = 0 \quad \Rightarrow \quad \mathcal{L}(\mathbf{x}) = f(\mathbf{x})$$

and

$$\inf_{\mathbf{x} \in \text{int}(N)} \mathcal{L}(\mathbf{x}) = \min_{\mathbf{x} \in N} f(\mathbf{x}) \quad \Leftrightarrow \quad (\mathbf{P})$$

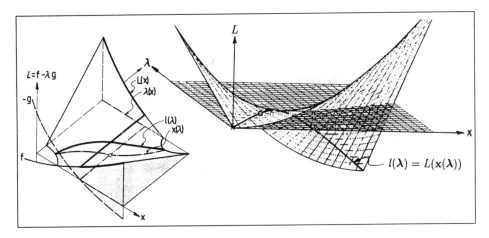

Figure 7: Saddle surface of the Lagrangian for a convex problem

The alternative approach via maximizing $l(\boldsymbol{\lambda})$ is much more promising because of the existence and uniqueness of the continuously differentiable relation $\mathbf{x}(\boldsymbol{\lambda})$. This can be proved by the *implicit function theorem* [6] applied to the stationarity $\frac{\partial L(\mathbf{x},\boldsymbol{\lambda})}{\partial \mathbf{x}} = 0$ using the non singularity of the Hessian matrix $\boldsymbol{D}_x^2 L$. In the case of differentiable functions $f, \mathbf{g}$ the constraint of (16) can be rewritten as stationarity condition $\boldsymbol{D}_x L(\mathbf{x},\boldsymbol{\lambda}) = 0$. The minimum for $\mathbf{x}$ is ensured by the positive definitness of the Hessian matrix and we shall refer to (16) as the *dual* of (**P**):

(**D**) $$\max_{\lambda \geq 0} L(\mathbf{x},\boldsymbol{\lambda}) \quad \text{s.t.} \quad \boldsymbol{D}_x L(\mathbf{x},\boldsymbol{\lambda}) = 0. \qquad (17)$$

In passing by the duality theorems, it is worthwile to note that the solution of the primal problem (6) is a solution of the dual problem (17) and vice versa if the problem is convex and regular. For the details of the duality theory see [Wolfe 61] and [Werner 84]. The weak duality theorem provides lower bounds on the objective function but for general nonlinear programming problems (NLP) there might be a gap between the solution of the dual problem and the primal problem. For convex NLP the strong duality theorem proves that the solutions are equal without this *duality gap*.

---

[6] see e.g. [Ort/Rhe 70]

7 Explicit approximations and sequential convex programming

7.1 CONLIN and MMA aproximation

Coming back to 'hill climbing' on the 'path' $\mathbf{x}(\boldsymbol{\lambda})$. A natural way would be to solve the dual constraints $\boldsymbol{D}_x L(\mathbf{x}, \boldsymbol{\lambda}) = \mathbf{0}$ for the explicit relation $\mathbf{x}(\boldsymbol{\lambda})$. With $\mathbf{x}(\boldsymbol{\lambda})$ on hand, backsubstitution into (17) would generate the *quasi unconstrained dual problem*:

$$\max_{\boldsymbol{\lambda} \geq \mathbf{0}} L(\mathbf{x}(\boldsymbol{\lambda}), \boldsymbol{\lambda}) = \max_{\boldsymbol{\lambda} \geq \mathbf{0}} l(\boldsymbol{\lambda}).$$

Most of the popular explicit approximations are addressed to yield separable Lagrangian functions, see [Fle/Bra 86] and [Svanb 87]. Now we will try to find an explicit relation of $\mathbf{x}(\boldsymbol{\lambda})$ by introducing a further mathematical property, the *separability*.

A problem is said to be *separabel* if f and $\mathbf{g}$ can be written as

$$f(\mathbf{x}) = \sum_{i=1}^{n} f_i(x_i) \quad \text{and} \quad g_j(\mathbf{x}) = \sum_{i=1}^{n} g_{ji}(x_i) \qquad j=1,\ldots,m. \qquad (18)$$

The separability of constraints as well as of the objective function generates a separabel Lagrangian function. Hence the solution of the dual constraints (17) can be done by many simple univariate steps instead of one multivariate *primal step*:

$$\begin{aligned}
\min_{\mathbf{x}} L(\mathbf{x}, \boldsymbol{\lambda}) &= \min_{\mathbf{x}} \left\{ f(\mathbf{x}) + \boldsymbol{\lambda}^T \mathbf{g}(\mathbf{x}) \right\} = \\
&= \min_{\mathbf{x}} \left\{ \sum_i f_i(x_i) + \sum_j \lambda_j \sum_i g_{ji}(x_i) \right\} = \\
&= \sum_i \min_{x_i} \left\{ f_i(x_i) + \sum_j \lambda_j g_{ji}(x_i) \right\} = \\
&= \sum_i \min_{x_i} L_i(x_i, \boldsymbol{\lambda}).
\end{aligned}$$

Now assume the original constraints and, if necessary, the objective function has been approximated by separable convex functions. This primal step for the approximated problem is easy to solve by n single minimization steps $\min_{x_i} L_i(x_i, \boldsymbol{\lambda})$. However, if the primal steps can be done analytically by evaluation of the stationarities $\boldsymbol{D}_{x_i} L_i(x_i, \boldsymbol{\lambda}) = 0$ for $x_i^* = x_i^*(\boldsymbol{\lambda})$ than we have the above mentioned quasi unconstrained *dual problem*. This problem can be solved for $\boldsymbol{\lambda}^*$ by any kind of mathematical programming methods for unconstrained optimization. At the solution $(\mathbf{x}^*, \boldsymbol{\lambda}^*)$ the problem will be approximated again, and a sequence of convex approximations (SCP) solves the original problem. This approach has been introduced by [Fleury 79] and [Svanb 82].

From the successful applied *Method of Moving Asymptotes* (MMA):

$$g_{MMA}(\mathbf{x}) = \tilde{g} - \sum_+ \tilde{g}_i(u_i - \tilde{x}_i) - \sum_- \tilde{g}_i(\tilde{x}_i - l_i) + \sum_+ \tilde{g}_i \frac{(u_i - \tilde{x}_i)^2}{u_i - x_i} + \sum_- \tilde{g}_i \frac{(\tilde{x}_i - l_i)^2}{x_i - l_i}$$

and after simple reformulation we get

$$g_{MMA}(\mathbf{x}) = \tilde{g} + \sum_+ \tilde{g}_i \frac{u_i(x_i - \tilde{x}_i) - \tilde{x}_i x_i + \tilde{x}_i^2}{u_i - x_i} + \sum_- \tilde{g}_i \frac{l_i(\tilde{x}_i - x_i) + \tilde{x}_i x_i - \tilde{x}_i^2}{x_i - l_i}$$

where $\tilde{g} = g(\tilde{\mathbf{x}})$, $\tilde{g}_i = \frac{\partial g(\tilde{\mathbf{x}})}{\partial x_i}$, and $\sum_{+/-}$ is the summation for positive/negative derivatives.

We can now derive some important standard approximations by passing the asymptotes to the limits:

- linear approximation:
  $\lim_{\substack{u_i \to +\infty \\ l_i \to -\infty}} g_{MMA} = g_L(\mathbf{x}) = \tilde{g} + \sum_i \tilde{g}_i(x_i - \tilde{x}_i)$
- reciprocal approximation:
  $\lim_{\substack{u_i \to 0 \\ l_i \to 0}} g_{MMA} = g_R(\mathbf{x}) = \tilde{g} + \sum_i \tilde{g}_i(x_i - \tilde{x}_i) \cdot \frac{\tilde{x}_i}{x_i}$
- CONLIN approximation:
  $\lim_{\substack{u_i \to +\infty \\ l_i \to 0}} g_{MMA} = g_C(\mathbf{x}) = \tilde{g} + \sum_+ \tilde{g}_i(x_i - \tilde{x}_i) + \sum_- \tilde{g}_i(x_i - \tilde{x}_i) \cdot \frac{\tilde{x}_i}{x_i}$

The latter approximation has been introduced – somewhat unsuitable termed as CONLIN = CONvex LINearisation – by [Fle/Bra 86] to handle the explicit dual formulation of structural optimization problems. A similar approximation called *conservative convex* or *hybrid* approximation has already been used by [Sta/Haf 79]. The essential properties of CONLIN and MMA are the *separability*, the *convexity* and the *ability to execute the primal phase explicitly*.

7.2 Motivation for the successful use of CONLIN and MMA with displacement related constraints

The evaluation of the implicit defined displacements are very CPU–time consuming. Therefore explicit approximations are used to speed up the optimization process. But these approximations have to be appropriate, at least for two points of view. On the one hand the approximation should be good enough for a large vicinity around the extension point to allow a big size of steplength for the optimization procedure. On the other hand the mathematical structure should hit the existing programming codes. By application of the *Cramer* rule as detailed below, we shall see that the elastic displacements can be shown to be proper fractional functions. Remember, the discussed nonlinear approximations are of this type. Although the *Cramer* rule is of no practical use for numerical calculations but for theoretical investigations it is a useful technique.

Consider once more the linear elements $x_i, i = 1, \ldots, n$, with their stiffness matrices

$$\mathbf{k}_i(\mathbf{x}) = x_i \mathbf{k}_i \quad \Rightarrow \quad \mathbf{K}(\mathbf{x}) = \sum_i x_i \mathbf{k}_i = \left(\sum_i k_{irs} \right)_{r,s \in D},$$

where $D = \{1, \cdots, d = DOF\}$. For the resulting elastic displacements $\mathbf{u}(\mathbf{x})$

$$\mathbf{u}(\mathbf{x}) = \mathbf{K}^{-1}(\mathbf{x})\mathbf{p} \quad \Rightarrow \quad u_r(\mathbf{x}) = \mathbf{e}_r^T \mathbf{K}^{-1}(\mathbf{x})\mathbf{p}$$

we need to have the inverse stiffnessmatrix

$$\mathbf{K}^{-1}(\mathbf{x}) = \left(k_{rs}^{-1}(\mathbf{x}) \right)_{r,s \in D} \qquad k_{rs}^{-1} = (-1)^{r+s} \frac{\det(\mathbf{K}_{rs})}{\det(\mathbf{K})},$$

where $\mathbf{K}_{rs}$ is the algebraic complement (adjoint) matrix of k_{rs}. With the determinants

$$\det(\mathbf{K}) = \sum_{\pi \in S(D)} sgn(\pi) \prod_{j=1}^{d} \sum_i x_i k_{ij\pi(j)}$$

$$\det(\mathbf{K}_{rs}) = \sum_{\pi \in \widehat{S}(D)} sgn(\pi) \prod_{j \neq r}^{d} \sum_i x_i k_{ij\pi(j)}$$

where $S(D)$ is the symmetric group of D and $\widehat{S}(D)$ are the bijective maps $\pi : D \setminus \{r\} \to D \setminus \{s\}$ or more convenient with the multi–index $\boldsymbol{\alpha} = (\alpha_1, \cdots, \alpha_n)$ of the order $\sigma \boldsymbol{\alpha} = \sum_i \alpha_i;\quad \mathbf{x}^{\boldsymbol{\alpha}} = \prod_i x_i^{\alpha_i}$

$$\det(\mathbf{K}) = \sum_{\sigma \alpha = d} a_\alpha \mathbf{x}^{\boldsymbol{\alpha}} \qquad \det(\mathbf{K}_{rs}) = \sum_{\sigma \alpha = d-1} b_\alpha \mathbf{x}^{\boldsymbol{\alpha}}$$

we get the selected displacement $u_r = \mathbf{e}_r^T \mathbf{K}^{-1} \mathbf{p}$

$$u_r(\mathbf{x}) = \sum_{k=1}^{d} k_{rk}^{-1} \cdot p_k = \sum_{k=1}^{d} (-1)^{r+k} \frac{\sum_{\sigma \alpha = d-1} b_\alpha \mathbf{x}^{\boldsymbol{\alpha}}}{\sum_{\sigma \alpha = d} a_\alpha \mathbf{x}^{\boldsymbol{\alpha}}} \cdot p_k$$

which is a homogeneous, proper fractional function as the variable term of the approximations discussed above:

$$\hat{u}(\mathbf{x}) = \sum_{k=1}^{n} \frac{\hat{q}_k}{x_k} = \sum_{k=1}^{n} \frac{\prod_{j \neq k}^{n} x_j}{\prod_{j=1}^{n} x_j} \cdot \hat{q}_k.$$

Generalizations are conceivable but the convexity might be lost:

$$\check{u}(\mathbf{x}) = \sum_{k=1}^{d} \frac{\check{q}_k}{\sum_i x_i b_{ik}} = \sum_{k=1}^{d} \frac{\prod_{j \neq k}^{d} \sum_i x_i b_{ij}}{\prod_{j=1}^{d} \sum_i x_i b_{ij}} \cdot \check{q}_k.$$

Remarks:

- In general only a few numbers of elements are joined to the same degree of freedom. Thus the approximation in use could be extended by taking into account more then one variable to the denominator.

- A first step forward to generalize the dual approach has been done by [Lootsma 89] with the introduction of two variables by means of partially–separability: $\hat{g}(\mathbf{x}) = \sum_i g_i(x_i, x_{i+1})$

8 Mathematical properties and how to exploit them

8.1 Duality with homogeneous functions

If the constraint functions $\mathbf{g}(\mathbf{x})$ and the objective function $f(\mathbf{x})$ of the *primal problem* (6) are *homogeneous* of the degree q and p respectively i.e. for $\alpha \in \mathbf{R}$ holds

$$\mathbf{g}(\alpha \cdot \mathbf{x}) = \alpha^q \cdot \mathbf{g}(\mathbf{x}) \quad \text{and} \quad f(\alpha \cdot \mathbf{x}) = \alpha^p \cdot f(\mathbf{x}) \tag{19}$$

then it follows from the Euler theorem:

$$q \cdot \mathbf{g}(\mathbf{x}) = \mathbf{D}_x \mathbf{g} \, \mathbf{x} \quad \text{and} \quad p \cdot f(\mathbf{x}) = \mathbf{D}_x f \, \mathbf{x}.$$

The constraint of the dual formulation (17) for continuously differentiable functions reads

$$\mathbf{D}_x f + \boldsymbol{\lambda}^T \mathbf{D}_x \mathbf{g} = \mathbf{0}$$

and multiplication from the right hand side by $\mathbf{x}$

$$D_x f\,\mathbf{x} + \boldsymbol{\lambda}^T D_x \mathbf{g}\,\mathbf{x} = 0$$

and making use of the homogeneity (19)

$$p \cdot f + q \cdot \boldsymbol{\lambda}^T \mathbf{g} = 0 \quad \Rightarrow \quad \boldsymbol{\lambda}^T \mathbf{g} = -\frac{p}{q} \cdot f \qquad (20)$$

the objective function of the dual (17) can be rewritten as

$$L(\mathbf{x}, \boldsymbol{\lambda}) = f + \boldsymbol{\lambda}^T(\mathbf{g} - \tilde{\mathbf{g}}) \stackrel{(20)}{=} f - \frac{p}{q} \cdot f - \boldsymbol{\lambda}^T \tilde{\mathbf{g}} = \left(\frac{q-p}{q}\right) \cdot f - \boldsymbol{\lambda}^T \tilde{\mathbf{g}}.$$

Application of this result to a given design $\mathbf{x}^\nu$ and solving the linear programming problem which is often called *linear dual*:

$$\max_{\boldsymbol{\lambda} \geq 0} \left(\frac{q-p}{q}\right) \cdot f(\mathbf{x}^\nu) - \boldsymbol{\lambda}^T \tilde{\mathbf{g}} \quad \text{s.t.} \quad D_x f(\mathbf{x}^\nu) + \boldsymbol{\lambda}^T D_x \mathbf{g}(\mathbf{x}^\nu) = 0 \qquad (21)$$

for $\boldsymbol{\lambda}^\nu(\mathbf{x}^\nu)$ yields by use of the duality theorem the greatest lower bound to the minimum $f(\mathbf{x}^*)$:

$$l(\boldsymbol{\lambda}^\nu) = L(\boldsymbol{\lambda}^\nu(\mathbf{x}^\nu)) \leq f(\mathbf{x}^*) \leq f(\mathbf{x}^\nu)$$

The *linear dual* (21) can be solved by the SIMPLEX algorithm. It is worth to do this at the first iteration steps because the result may be useful for two reasons. First, $\boldsymbol{\lambda}^\nu(\mathbf{x}^\nu)$ is in general a good estimate for the dual variables, which are the Lagrangian multipliers. These variables can be used to start any kind of dual algorithms as well as the codes of *optimality criteria*. Second, the result $L(\boldsymbol{\lambda}^\nu(\mathbf{x}^\nu))$ is a lower bound for $f(\mathbf{x}^*)$ and the relative distance to the upper bound $f(\mathbf{x}^\nu)$ might be used to monitor the process by the *termination criterion*:

$$l(\boldsymbol{\lambda}^\nu) > \rho \cdot f(\mathbf{x}^\nu) \qquad \text{where} \quad \rho < 1.$$

Note that the bound $l(\boldsymbol{\lambda}^\nu)$ is dependent on the current design $\mathbf{x}^\nu$ thus $l(\boldsymbol{\lambda}^\nu) = f(\mathbf{x}^\nu)$ is only true for $\mathbf{x}^\nu = \mathbf{x}^*$. The value ρ should be seen as *design optimality*.

Now we shall show that homogeneity is a relevant property for the problem in question i.e. displacement related constraints. If for the discrete idealized FEM–structure elements from the same degree s are implemented only

$$\mathbf{k}_i(x_i) = x_i^s \cdot \mathbf{k}_i \quad \text{for any } i = 1, \ldots, n \quad \Rightarrow$$

$$\mathbf{K}(\alpha \cdot \mathbf{x}) = \sum_i \alpha^s \cdot x_i^s \cdot \mathbf{k}_i = \alpha^s \cdot \mathbf{K}(\mathbf{x}) \Rightarrow \mathbf{K}^{-1}(\alpha \cdot \mathbf{x}) = \alpha^{-s}\mathbf{K}^{-1}(\mathbf{x})$$

$$\text{and} \quad \mathbf{u}(\alpha \cdot \mathbf{x}) = \mathbf{K}^{-1}(\alpha \cdot \mathbf{x})\mathbf{p} = \alpha^{-s}\mathbf{K}^{-1}(\mathbf{x})\mathbf{p} = \alpha^{-s}\mathbf{u}(\mathbf{x}).$$

The homogeneity (19) holds for stress or equivalent stress with $\mathbf{S}$ and $\mathbf{V}$ as constant matrices

$$\sigma(\alpha \cdot \mathbf{x}) = \mathbf{S}\mathbf{u}(\alpha \cdot \mathbf{x}) = \alpha^{-s} \cdot \sigma(\mathbf{x})$$

$$\sigma_V(\alpha \cdot \mathbf{x}) = \sigma(\alpha \cdot \mathbf{x})^T \mathbf{V} \sigma(\alpha \cdot \mathbf{x}) = \alpha^{-2s} \cdot \sigma_V(\mathbf{x}).$$

Finally it should be mentioned that *scaling to the feasible domain* is also based on homogeneity:

Consider the half-line $\mathbf{x}(\alpha) = \alpha \cdot \mathbf{x}^\nu$ for a given design $\mathbf{x}^\nu$. We are looking for a scalar $\alpha \in \mathbf{R}$ that scales the design $\tilde{\mathbf{x}} \stackrel{\text{def}}{=} \alpha \cdot \mathbf{x}^\nu$ to the boundary of the feasible domain i.e. $g(\tilde{\mathbf{x}}) = \tilde{g}$:

$$g(\alpha \cdot \mathbf{x}^\nu) \stackrel{(19)}{=} \alpha^q \cdot g(\mathbf{x}^\nu) \stackrel{!}{=} \tilde{g} \Rightarrow$$

$$\tilde{\mathbf{x}} = \alpha \cdot \mathbf{x}^\nu = \left(\frac{\tilde{g}}{g(\mathbf{x}^\nu)}\right)^{\frac{1}{q}} \mathbf{x}^\nu.$$

Applied to the linear dual (21) it can be shown that

$$\boldsymbol{\lambda}(\alpha) = \boldsymbol{\lambda}(\alpha \cdot \mathbf{x}^\nu) = \alpha^{p-q} \cdot \boldsymbol{\lambda}(\mathbf{x}^\nu)$$

is a homogeneous function of degree $p - q$ and the constraint (20) of the linear dual (21) holds for any $\alpha \in \mathbf{R}$. Hence, the lower bound

$$l(\boldsymbol{\lambda}(\alpha)) = \left(\frac{q-p}{q}\right) \alpha^p f^\nu - \alpha^{p-q} \boldsymbol{\lambda}^T(\mathbf{x}^\nu)\tilde{\mathbf{g}}$$

might be improved by maximizing with respect to the scaling variable $\alpha \in \mathbf{R}$. With the stationarity $\frac{dL}{d\alpha} = 0$ and the maximum condition $\frac{d^2 L}{d\alpha^2} < 0$ we get for $\text{sgn}(p) \neq \text{sgn}(q)$:

$$l(\tilde{\alpha}) = \left(\frac{r \cdot f^\nu}{\boldsymbol{\lambda}^T \tilde{\mathbf{g}}}\right)^r f^\nu \quad \text{and} \quad \tilde{\alpha} = \left(\frac{\boldsymbol{\lambda}^T \tilde{\mathbf{g}}}{r \cdot f^\nu}\right)^{\frac{1}{q}} \quad \text{where} \quad r = -\frac{p}{q} > 0.$$

This result for $p = 1$, $q = -1$ has already been used by [Barthol 79].

8.2 Bilinearity for displacement constraints and the integrated formulation

Up to now we have only considered sizing variables $x_i, i = 1, \ldots, n$. Why not taken the displacements $u_j, j = 1, \ldots, d$ (d=degree of freedom) as design variables as well? From the system equations of elastic equilibrium we will get as many constraints as additional variables. The implicit defined displacement inequality constraints will now degenerate to simple side constraint. The mathematical formulation of this problem

$$\min_x f(\mathbf{x}) = \mathbf{c}^T \mathbf{x}$$
$$\text{s.t.} \quad u_{lj} \leq u_j(\mathbf{x}) = \mathbf{e}_j^T \mathbf{K}^{-1}(\mathbf{x})\mathbf{p} \leq u_{uj} \quad j \in D$$
$$x_{li} \leq x_i \leq x_{ui} \quad i = 1, \ldots, n$$

is equivalent to the *integrated formulation*:

$$\min_x f(\mathbf{x}) = \mathbf{c}^T \mathbf{x}$$
$$\text{s.t.} \quad g_j(\mathbf{x}, \mathbf{u}) \stackrel{\text{def}}{=} \mathbf{e}_j^T \mathbf{K}(\mathbf{x})\mathbf{u} = p_j \quad j = 1, \ldots, d$$
$$x_{li} \leq x_i \leq x_{ui} \quad i = 1, \ldots, n$$
$$u_{lj} \leq u_j \leq u_{uj} \quad j = 1, \ldots, d.$$

Proposition: The equality constraints $g_j(\mathbf{x}, \mathbf{u}), j \in D$ for the integrated formulation are *bilinear* if only linear elements are used to build up the FE model.

Proof:

$$g_j(\mathbf{x}, \mathbf{u}) = \mathbf{e}_j^T \mathbf{K}(\mathbf{x})\mathbf{u} = \sum_{k=1}^{d} \left(\sum_{i=1}^{n} x_i \cdot k_{jk}^{(i)} \right) u_k$$
$$= \sum_{i=1}^{n} x_i \left(\sum_{k=1}^{d} k_{jk}^{(i)} \cdot u_k \right) = \left(\sum_{i=1}^{n} x_i \left\{ k_{jk}^{(i)} \right\}_{k=1,\ldots,d} \right) \mathbf{u}$$
$$= \mathbf{x}^T \left[k_{jk}^{(i)} \right]_{\substack{i=1,\ldots,n \\ k=1,\ldots,d}} \mathbf{u} = \mathbf{x}^T \mathbf{A}_j \mathbf{u} \quad \square$$

Now we have to assess the pros and cons of this formulation. The advantage is the simple mathematical structure of the constraints. It is a question to the mathematical programming people whether they can make use of the bilinearity in optimization codes or not. The disadvantage of the integrated formulation is obviously the large number of variables as well as of constraints. If there are special algorithms to deal with bilinear equality constraints we should check which formulation is more effective.

References

[Barthol 79]   P. Bartholomew: *A dual bound used for monitoring structural optimization programs*, J. Engineering Optimization, Vol. 4, pp. 45–50, 1979.

[Bel/Aro 84] A.D. Belegundu and J.S. Arora: *A computational study of transformation methods for optimal design*, AIAA J., Vol. 22, No. 4, pp. 535–542, 1984.

[Bel/Aro 85] A.D. Belegundu and J.S. Arora: *A study of mathematical programming methods for structural optimization. Part I: Theory / Part II: Numerical results*, Int. Jour. Num. Meth. Engrg., Vol. 21, pp. 1583–1599 / pp. 1601–1623 , 1985.

[Ben/Kik 88] M.P. Bendsøe and N. Kikuchi: *Generating optimal topologies in structural design using a homogenization method*, Comp. Meths. Appl. Mech. Engrg., Vol. 71, pp. 197–224, 1988.

[Bendsøe 89] M.P. Bendsøe: *Optimal shape design as a material distribution problem*, Structural Optimization, Vol. 1(4), pp. 193–202, 1989.

[Bendsøe 92] M.P. Bendsøe: *Optimisation topologique*, Lecture Notes, COMETT Seminar on Conception optimale de structures assistée par ordinateur, Nice, France, July 13–17, 1992, writen in english.

[Ber/Ven 84] L. Berke and V.B. Venkayya: *Review of optimality criteria approaches to structural optimization*, ASME Struct. Opt. Symp., pp. 22–34, 1984.

[Bot/Ben 84] M.E. Botkin and J.A. Bennett: *Shape optimization of three- dimensional folded plate structures*, AIAA J., Vol. 23, No. 11, pp. 1804–1810, 1984.

[Carmich 81] D.G. Carmichael: Chapter 10 in *Structural modelling and optimization a general methodology for engineering and control*, Ellis Horwood Ltd., Series in Civil Engineering, 1981.

[Dorn et al. 64] W.S. Dorn, R.E. Gomory and H.J. Greenberg: *Automated design of optimal structures*, Jur. de Mecanique, Vol. 3, No. 1, pp. 25–52, 1964.

[Esch et al. 85] H. Eschenauer et al.:*Rechnerische und experimentelle Untersuchungen zur Strukturoptimierung von Bauweisen*, Forschungsbericht der DFG des Inst. f"ur Mechanik und Regelungstechnik der Univ. Ges. Hochschule Siegen, pp. 9–12,1985.

[Esping 83] B.J.D. Esping:*CAD Approach to the minimum weight design problem*, Report 83–14, Royal Inst. of Techn. Stockholm, Sweden, 1983.

[Fleury 79] C. Fleury:*Structural weight optimization by dual methods of convex programming*, Int. Jour. Num. Meth. Engrg., Vol. 14, pp. 1761–1783, 1979.

[Fleury 85] C. Fleury:*Shape optimal design by the convex linearization method*, Int. Symp. *The optimum shape: Automated structural design*, Warren, Michigan.

[Fleury 86] C. Fleury: *Computer aided optimal design of elstic structures; Part II: Convex approximation strategies in structural synthesis*, in Computer Aided Optimal Design: Structural and Mechanical Systems, edited by C.A. Mota Soares, NATO ASI, Series F, Vol. 27, Springer—Verlag 1986, ISBN 3-540-17598-9.

[Fleury 89a]   C. Fleury:*First and second order convex approximation strategies in structural optimization*, Structural Optimization, Vol. 1(1), pp. 1–10, 1989.

[Fleury 89b]   C. Fleury:*CONLIN: an efficient dual optimizer based on convex approximation concepts*, Structural Optimization, Vol. 1(2), pp. 81–89, 1989.

[Fle/Bra 86]   C. Fleury and V. Braibant:*Structural optimization — a new dual method using mixed variables*, Int. Jour. Num. Meth. Engrg., Vol. 23, pp. 409–428, 1986.

[Haf/Ade 89]   R.T. Haftka and H.M. Adelman: *Recent developments in structural sensitivity analysis*, Structural Optimization, Vol. 1(3), pp. 137–151, 1989.

[Haf/Bar 91]   R.T. Haftka and B. Barthelemy: *On the accuracy of shape sensitivity*, Structural Optimization, Vol. 3(1), pp. 1–6, 1991.

[Hemp 64]   W.S. Hemp: *Studies in the theory of Michell structures*, Proc. Int. Congr., Appl. Mech., M"unchen, 1964.

[Hug et al. 80]   O.F. Hughes, F. Mistree and V. Zanic : *A practical method for rational design of ship structures*, J. Ship Res., Vol. 24, No. 2, pp. 101–113, 1980.

[Kik/Suz 91]   N. Kikuchi and K. Suzuki: *Structural optimization of a linearly elastic structure using the homogenization method*, In: Composite Media and Homogenization Theory, edited by G. Dal Masso and C.F. Dell Antonio, pp. 183–204, Birkh"auser Verlag, Boston, 1991.

[Lootsma 89]   F.A. Lootsma: *A comparative study of primal and dual approaches for separable and partially–separable nonlinear optimization problems*, Structural Optimization, Vol. 1(2), pp. 73–79, 1989.

[Maxwe 1869]   J.C. Maxwell: *Scientific Papers*, Vol. 2, p. 175, Cambridge Univ. Press, 1869.

[Miche 1904]   A.G.M Michell: *The limits of economy of material in frame structures*, Philosophical Magazine, Series 6, Vol. 8, pp. 589–597, 1904.

[Ort/Rhe 70]   J.M. Ortega and W.C. Rheinbold: *Iterative solution of nonlinear equations in several variables*, ACADEMIC PRESS, 1970.

[Peder 73]   P. Pedersen: *Optimal joint positions for space trusses*, Jour. of Struc. Div. ASCE, Vol. 99, No. 12, pp. 2459–2476, 1973.

[Prasad 83]   B. Prasad: *Potential forms of explicit constraint approximations in structural optimization - Part 1: Analysis and projections / Part 2: Numerical experience*, Comp. Meth. Appl. Mech. Engrg., Vol. 40, pp. 1–26 /, Vol. 46, pp. 15–38 , 1983.

[San/Rag 80]   E. Sandgren and K.M. Ragsdell: *The utility of nonlinear algorithms: A comparative study - Part 1 and Part 2*, ASME J. Mech. Des., Vol. 102(3), pp. 540–551, 1980.

[Schittk 80]   K. Schittkowski: *Nonlinear programming codes. Information, test, performance*, Lecture Notes in Economics and Mathematical Systems, Springer Verlag, 1980.

[Schm/Far 74]   L.A. Schmit and B. Farshi: *Some approximation concepts for structural sythesis*, AIAA Jour., Vol. 12, No. 5, pp. 692–699, 1974.

[Schm/Can 84]   L.A. Schmit and K.J. Chang: *A multi-level method for structural synthesis*, 25th AIAA/ASME/ASCE/AHS Structures, Structural Dynamics and Material Conf., Palm Springs, 1984.

[Sob et al. 85]   J. Sobieski, B.B. James and M.F. Riley: *Structural optimization by generalized multilevel optimization*, 26th AIAA/ASME/ASCE/AHS Structures, Structural Dynamics and Material Conf., Orlando, Florida, 1984.

[Stadler 84]   W. Stadler: *Multicriteria optimization in mechanics –A survey–*, Appl. Mech. Review, Vol. 37, No. 3, pp. 277–286, 1984.

[Sta/Haf 79]   J.H.Jr. Starnes and R.T. Haftka: *Preliminary design of wings for buckling, stress and displacement constraints*, Jour. Aircrafts, Vol. 16, pp. 564–570, 1979.

[Stoer 63]   J. Stoer: *Duality in nonlinear programming and the minimax theorem*, Numer. Math., Vol. 5, pp. 371–379,

[Suz/Kik 91]   K. Suzuki and N. Kikuchi: *Layout optimization using the homogenization method: Generalized layout design of three dimensional shells for car bodies*, In: Optimization of large structural systems, edited by G.I.N. Rozvany, Lecture Notes, NATO ASI, Berchtesgaden, Vol. 3, pp. 110–126, FRG, 1991.

[Svanb 82]   K. Svanberg: *An algorithm for optimum structural design using duality*, Math. Programming Study, Vol. 20, pp. 161–177, 1982.

[Svanb 84]   K. Svanberg: *On local and global minima in structural optimization*, New Directions in Optimum Structural Design, edited by Atrek, Gallagher, Ragsdell and Zienkiewicz, John Wiley & Sons, 1984, ISBN 0-471-90291-8, pp. 327–341.

[Svanb 87]   K. Svanberg: *Method of moving asymptotes — a new method for structural optimization*, Int. Jour. Num. Meth. Engrg., Vol. 24, pp. 359–373, 1987.

[Van/Mos 72]   G.N. Vanderplaats and F. Moses: *Automated design of trusses for optimum geometry*, Jour. of Struc. Div. ASCE, Vol. 89, No. 6, pp. 671–690, 1972.

[Werner 84]   J. Werner: *Optimization theory and applications*, Vieweg Advanced Lectures in Mathematics, Verlag Friedr. Vieweg & Sohn,ISBN 3-528-08594-0.

[Wolfe 61]   P. Wolfe: *A duality theorem for non–linear programming*, Quart. Appl. Math., Vol. 19, pp. 239–244, 1961.

Mathematical Optimization
- An Introduction -

Klaus Schittkowski

Abstract. The intention of the article is to provide some mathematical background to understand at least the basic ideas behind the optimization algorithms implemented in structural optimization systems. For the smooth constrained nonlinear programming problem, a brief introduction into some basic theoretical concepts is given, i.e. the necessary and sufficient optimality criteria are discussed. They are required to understand the solution approaches for solving structural mechanical optimization problems and to analyse the results achieved. In particular the sequential linear and quadratic programming and the convex approximation methods are described briefly. They represent the most frequently used numerical algorithms of structural optimization.

1. The nonlinear programming problem

The design of a mechanical structure is often based on the requirement to *optimize* a suitable criterion in the sense to get a better design according to the criterion chosen, and to retain feasibility subject to the constraints that must be satisfied. The more complex the structure is, the more difficult is the empirical iterative refinement *by hand* based on successive analysis.

In the subsequent articles of this book, several structural optimization software systems will be introduced which are based on a finite element analysis. In all cases, the underlying mechanical design problem can be modelled and described in abstract terms, so that a mathematical nonlinear programming problem of the following form is formulated:

$$x \in \mathsf{R}^n : \begin{array}{ll} \min f(x) & \\ g_j(x) = 0 & , j = 1, ..., m_e \\ g_j(x) \geq 0 & , j = m_e + 1, ..., m \\ x_l \leq x \leq x_u & \end{array} \qquad (1)$$

We may imagine, for example, that the objective function describes the weight of a structure that is to be minimized subject to sizing and shape variables, and that the constraints impose limitations on structural response quantities, e.g. upper bounds for stresses or displacements under static loads. Many other objectives or constraints can be modelled in a way so that they fit into the above general frame.

Since, however, the mathematical theory and also the basic properties of the implemented numerical algorithms are well known, see e.g. Papalambros and Wilde (1988), we want to present here only a very brief review on the most important topics from the viewpoint of mathematical optimization. We proceed from the general formulation (1) without any further specialization of the model functions. It is assumed that the functions are continuously differentiable on the whole R^n. To simplify the subsequent notation, we omit the lower and upper bounds on the variables in the subsequent sections.

To understand the optimization algorithms, i.e. their mathematical structure and the results achieved, a brief introduction into the mathematical programming theory is presented in Section 2. It is sufficient to consider only the optimality criteria which contain all information to analyse a practical numerical method. Moreover the investigation of the optimality criteria helps to understand the results, e.g. to learn whether an achieved answer can be accepted or not or to get an idea on the final accuracy and the sensitivity of the solution.

The basic strategy of any optimization method is to approximate the given nonlinear problem by another problem that can be solved then much easier either directly or iteratively. One possible attempt leads to successive generation of linear programming problems which must be solved the by standard techniques. In this case, the resulting algorithm is called sequential linear programming or SLP-method and is described in Section 3.

In another approach, the original problem is approximated quadratically leading to so-called sequential quadratic programming or SQP-methods, as outlined in Section 4. These algorithms are general purpose methods, i.e. can be applied to a large variety of optimization problems from completely different disciplines.

In structural mechanical optimization, also algorithms based on convex nonlinear approximations of the original problem were developed in the past and work very successfully in practice at least under certain conditions that are often satisfied. Section 5 contains a description of this approach.

The three types of numerical algorithms discussed in this introduction, represent the most frequently used mathematical methods to solve structural design optimization problems. Thus most of the software systems described in this book, use the one or other method internally. However we must be aware that there exists a large variety of different nonlinear programming algorithms and variants of known ones, so that also optimization methods are mentioned in the subsequent papers which are not discussed in the introduction. In these cases the reader is refered to the literature to get more information.

2. Optimality criteria

It is outside the scope of this paper to present a detailed introduction into the mathematical theory behind the nonlinear programming problem (1). To get at least an impression on the basic ideas behind optimization algorithms and to learn the most important terms, a brief description of the mathematical optimality criteria is given. We need these criteria to understand the optimization algorithms and to be able to analyse their results.

First we have to specify the notation used in this paper. We denote by

$$\nabla f(x) := (\frac{\partial}{\partial x_1}f(x), ..., \frac{\partial}{\partial x_n}f(x))^T$$

the gradient of a differentiable function $f(x)$ for $x \in \mathsf{R}^n$. In case of doubt, we add the index x to ∇ to indicate that the differentiation is to be performed only with respect to the x-variables. Moreover, we write the Hessian matrix of a twice differentiable function $f(x)$ with respect to $x \in \mathsf{R}^n$ in the form

$$\nabla^2 f(x) := \left(\frac{\partial^2}{\partial x_i \partial x_j} f(x)\right)$$

Again an index x may insure that we consider only the differentiation with respect to x.

Now we define by

$$I(x) := \{j : g_j(x) = 0, m_e < j \leq m\}$$

the set of active constraints for any feasible x, i.e. any x that satisfies all restrictions.

The most important tool to understand the optimality criteria, is the so-called Lagrange function

$$L(x, u) := f(x) - \sum_{j=1}^{m} u_j g_j(x)$$

which is defined for $x \in \mathsf{R}^n$ and $u = (u_1, ..., u_m)^T$, and which describes a linear combination of the objective function and the constraints. The coefficients $u_j, j = 1, ..., m$, are called the Lagrange multipliers of problem (1).

Now we are able to formulate the optimality criteria. Since the required assumptions differ, we distinguish as usual between necessary and sufficient conditions. In the first case, we need an assumption called *constrained qualification* which means that for a feasible x, the gradients of active constraints, i.e. the set $\{\nabla g_j(x) : j \in I(x)\}$, are linearly independent.

Theorem: *Let f and g_j for $j = 1, .., m$ be twice contionuously differentiable functions, x^* be a local minimizer of (1) and the constrained qualification be satisfied in x^*. Then* there is a $u^* \in R^m$ so that the following conditions are satisfied:

a) $u_j^* \geq 0$ for $j = m_e + 1, ..., m$,
$\nabla_x L(x^*, u^*) = 0$,
$u_j^* g_j(x^*) = 0$ for $j = m_e + 1, ..., m$,

b) $s^T \nabla_x^2 L(x^*, u^*) s \geq 0$ for all $s \in R^n$ with $\nabla g_j(x^*)^T s = 0$,
$j \in \{1, ..., m_e\} \cup I(x^*)$

It is possible to show the opposite direction without the regularity condition *constraint qualification* subject to some stronger statement, i.e. a strict local minimizer.

Theorem: *Let f and g_j for $j = 1, .., m$ be twice contionuously differentiable functions, $x^* \in R^n$ be feasible with respect to (1) and $u^* \in R^n$ a multiplier vector with*

a) $u_j^* \geq 0$ for $j = m_e + 1, ..., m$,
$\nabla_x L(x^*, u^*) = 0$,
$u_j^* g_j(x^*) = 0$ for $j = m_e + 1, ..., m$,

b) $s^T \nabla_x^2 L(x^*, u^*) s > 0$ for all $s \in R^n, s \neq 0$, with $\nabla g_j(x^*)^T s = 0$,
$j \in \{1, ..., m_e\}$ and $j = m_e + 1, ..., m$ with $u_j^* > 0$.

Then x^* is an isolated local minimizer of f, i.e. there is a neighbourhood $U(x^*)$ of x^*, so that $f(x^*) < f(x)$ for all $x \in U(x^*)$, $x \neq x^*$.

The condition, that the gradient of the Lagrange function vanishes at an optimal solution, is called the *Kuhn-Tucker-condition* of (1). In other words, the gradient of f is a linear combination of gradients of active constraints:

$$\nabla f(x^*) = \sum_{i=1}^{m} u_j^* \nabla g_j(x^*)$$

The complementary slackness condition $u_j^* g_j(x^*) = 0$ together with the feasibility of x^* guarantees, that only the active constraints, i.e. the interesting ones, contribute a gradient in the above sum. Either a constraint is satisfied by equality or the corresponding multiplier value is zero.

The Kuhn-Tucker condition can be computed within an optimization algorithm, if suitable multiplier estimates are available, and serves as a stopping condition. However the second order condition b) can only be evaluated numerically, if second derivatives are available. The condition is required in the optimality criteria, to be able to distinguish between a stationary point and a local minimizer.

Proofs of the theorems are found in any textbook on nonlinear programming, e.g. in Gill, Murray and Wright (1981). An elementary outline of the optimality conditions and a geometric interpretation is found in Papalambros and Wilde (1988).

We should note here that there exist a couple of algorithms for the optimal design of structures, that are based on the above optimality conditions and which are called therefore the *optimality criteria methods* in engineering sciences.

3. Sequential linear programming methods

In particular for design optimization, sequential linear programming or SLP methods, respectively, are quite powerful due to the special problem structure and, in particular, numerical limitations that prevent the usage of higher order methods in some cases. The idea is to approximate the nonlinear problem (1) by linear one to get a new iterate. Thus the next iterate $x_{k+1} = x_k + d_k$ is formulated with respect to solution d_k of the following linear programming problem:

$$d \in \mathsf{R}^n : \begin{array}{l} \min \nabla f(x_k)^T d \\ \nabla g_j(x_k)^T d + g_j(x_k) = 0 \quad , j = 1, ..., m_e \\ \nabla g_j(x_k)^T d + g_j(x_k) \geq 0 \quad , j = m_e + 1, ..., m \\ \| d \| \leq \delta_k \end{array} \qquad (2)$$

The principle advantage is that the above problem can be solved by any standard linear programming software. Additional bounds for the computation of d_k are required to avoid bad estimates particularly in the beginning of the algorithm, when the linearization is too inaccurate. The bound δ_k has to be adapted during the algorithm. One possible way is to consider the so-called exact penalty function

$$p(x, r) := f(x) + \sum_{i=1}^{m_e} r_j |g_j(x)| + \sum_{i=m_e+1}^{m} r_j |\min(0, g_j(x))|$$

defined for each $x \in \mathsf{R}^n$ and $r = (r_1, ..., r_m)^T$. Moreover we need its first order Taylor approximation given by

$$\begin{array}{rl} p_a(x, d, r) := & f(x) + \nabla f(x)^T d + \sum_{i=1}^{m_e} r_j |g_j(x) + \nabla g_j(x)^T d| \\ & + \sum_{i=m_e+1}^{m} r_j |\min(0, g_j(x) + \nabla g_j(x)^T d)| \end{array}$$

Then we consider the quotient of the actual and predicted change at an iterate x_k and a solution d_k of the linear programming subproblem

$$q_k := \frac{p(x_k, r) - p(x_k + d_k, r)}{p(x_k, r) - p_a(x_k, d_k, r)}$$

where the penalty parameters are predetermined and must be sufficiently large, e.g. larger than the expected multiplier values at an optimal solution. The δ_k-update is then performed by

$$\delta_{k+1} := \begin{cases} \delta_k / \sigma & , \text{ if } q_k < \rho_1 \\ \delta_k \sigma & , \text{ if } q_k > \rho_2 \\ \delta_k & , \text{ otherwise} \end{cases}$$

Here $\sigma > 1$ and $0 < \rho_1 < \rho_2 < 1$ are constant numbers. Some additional safeguards are necessary to be able to prove convergence, confer Lasdon, Kim and Zhang (1983) or Fletcher and de la Maza (1987), for example.

4. Sequential quadratic programming methods

Sequential quadratic programming or SQP methods are the standard general purpose algorithms for solving smooth nonlinear optimization problems under the following assumptions:

- The problem is not too big.
- The functions and gradients can be evaluated with sufficiently high precision.
- The problem is smooth and well-scaled.

The mathematical convergence and the numerical performance properties of SQP methods are very well understood now and are published in so many papers, that only a few can be mentioned here. Theoretical convergence is investigated in Han (1976,1977), Powell (1978a,1978b), Schittkowski (1983), e.g., and the numerical comparative studies of Schittkowski (1980) and Hock, Schittkowski (1981) show their superiority over other mathematical programming algorithms under the above assumptions.

The key idea is to approximate also second order information to get a fast final convergence speed. Thus we define a quadratic approximation of the Lagrange function $L(x,u)$ and an approximation of the Hessian matrix $\nabla_x^2 L(x_k, u_k)$ by a so-called quasi-Newton matrix B_k. Then we get the subproblem

$$d \in \mathbb{R}^n : \begin{array}{l} \min \frac{1}{2} d^T B_k d + \nabla f(x_k)^T d \\ \nabla g_j(x_k)^T d + g_j(x_k) = 0 \quad , j = 1, ..., m_e \\ \nabla g_j(x_k)^T d + g_j(x_k) \geq 0 \quad , j = m_e + 1, ..., m \end{array} \qquad (3)$$

Instead of trust regions or move limits, respectively, as for SLP methods, the convergence is ensured by performing a line search, i.e. a steplength computation to accept a new iterate $x_{k+1} := x_k + \alpha_k d_k$ for an $\alpha_k \in (0,1]$ only if x_{k+1} satisfies a descent property with respect to a solution d_k of (3). Following the approach of Schittkowski (1983), e.g., we need also a simultaneous line search with respect to the multiplier approximations called v_k and define $v_{k+1} := v_k + \alpha_k(u_k - v_k)$ where u_k denotes the optimal Lagrange multiplier of the quadratic programming subproblem (3).

The line search is performed with respect to a merit function

$$\psi_k(\alpha) := \phi_{r_k}(x_k + \alpha d_k, v_k + \alpha(u_k - v_k))$$

and

$$\phi_r(x,v) := f(x) - \sum_{j=1}^{m_e}(v_j g_j(x) - \tfrac{1}{2} r_j g_j(x)^2) \\ - \sum_{j=m_e+1}^{m} \begin{cases} (v_j g_j(x) - \tfrac{1}{2} r_j g_j(x)^2), & \text{if } g_j(x) \leq v_j/r_j \\ \tfrac{1}{2} v_j^2 / r_j, & \text{otherwise} \end{cases}$$

where $r = (r_1, ..., r_m)^T$. We should note here that also other concepts, i.e. other merit functions are found in the literature. Then we initiate a subiteration starting with

$\alpha = 1$ and perform a successive reduction combined with a quadratic interpolation of $\psi_k(\alpha)$, until for the first time, a the stopping condition of the form

$$\psi_k(\alpha) \leq \psi_k(0) + \mu\alpha\psi'_k(0)$$

is satisfied, where we must be sure that $\psi'_k(0) < 0$, of course. To guarantee this condition, the penalty parameter r_k must be evaluated by a special formula which is not repeated here.

The update of the matrix B_k can be performed by standard techniques known from unconstrained optimization. In most cases, the BFGS-method is applied, a numerically simple rank-2-correction starting from the identy or any other positive definite matrix. Only the differences $x_{k+1} - x_k$, $\nabla_x L(x_{k+1}, u_k) - \nabla_x L(x_k, u_k)$ are required. Under some safeguards it is possible to guarantee that all matrices B_k are positive definite.

Among the most attractive features of SQP methods is the superlinear convergence speed in the neighbourhood of a solution given by

$$\| x_{k+1} - x^* \| \leq \gamma_k \| x_k - x^* \|$$

where γ_k is a sequence of positive numbers converging to zero and x^* an optimal solution.

To understand this convergence behavior, replace B_k by the true Hessian of the Lagrangian function and consider only equality constraints. Then it is very easy to see that an SQP method is nothing else than Newton's method for solving the nonlinear system of $n + m$ equations in $n + m$ unknowns given by the Kuhn-Tucker conditions. This result can be extended to inequality constraints as well. Then we get immediately the quadratic convergence behavior and, if we replace B_k again by its approximation, the weaker superlinear convergence rate.

5. Convex approximation methods

Convex approximation methods were developed in particular by Fleury (1986), and Svanberg (1987) extended his approach. Their key motivation was to implement an algorithm, which is particularly designed for solving mechanical structural optimization problems. Thus their domain of application is somewhat restrcicted to a special problem type.

Although equality constraints can be handled in the same way as for SLP and SQP methods, most convex approximation or CA methods, respectively, assume that only inequalities exist in the optimization problem.

The key idea is to use a convex approximation of the original problem (1) instead of a linear or quadratic one, and then to solve the resulting nonlinear subproblem by a specifically designed algorithm, that takes advantage of the simplified problem structure. Consequently CA methods are only useful in cases where the evaluation of function and gradient values is much more expensive than the internal computations to solve the subproblem.

Let us consider e.g. the objective function $f(x)$. By inverting suitable variables, we get the convex approximation of $f(x)$ in the neighbourhood of an $x_k \in \mathbb{R}^n$ by

$$f_k(x) := f(x_k) + \sum_{i \in I_k^+} \frac{\partial}{\partial x_i} f(x_k)(x_i - x_i^k) - \sum_{i \in I_k^-} \frac{\partial}{\partial x_i} f(x_k)(1/x_i - 1/x_i^k)(x_i^k)^2$$

where $x = (x_1, ..., x_n)^T$ and $x_k = (x_1^k, ..., x_n^k)^T$ and where

$$I_k^- := \{i : 1 \leq i \leq n, \frac{\partial}{\partial x_i} f(x_k) \leq 0\}$$

$$I_k^+ := \{i : 1 \leq i \leq n, \frac{\partial}{\partial x_i} f(x_k) > 0\}$$

The reason for inverting design variables in the above way is, that stresses and displacements are exact linear functions of the reciprocal sizing variables in case of a statically determined structure. Moreover the numerical experience shows that also in other cases, convex linearization is applied quite successfully in practice, in particular in shape optimization, although a mathematical motivation cannot be given in this case.

In a similar way, reciprocal variables are introduced for the inequality constraints, where we have to change the signs to get a concave function approximation, but, on the other hand, a convex feasible region of the subproblem. The corresponding index sets are denoted by $I_{k,j}^+$ and $I_{k,j}^-$ for $j = 1, ..., m$.

After some reorganization of constant data, we get a convex subproblem of the following form:

$$x \in \mathbb{R}^n : \begin{array}{l} \min \sum_{i \in I_k^+} f_i^k x_i - \sum_{i \in I_k^-} f_i^k / x_i \\ \sum_{i \in I_{k,j}^-} g_{i,j}^k x_i - \sum_{i \in I_{k,j}^+} g_{i,j}^k / x_i + \bar{g}_{i,j}^k \geq 0, \quad j = 1, ..., m \end{array} \quad (4)$$

The solution of the above problem determines then the next iterate x_{k+1}. We do not investigate here the question how the mathematical structure of (4) can be exploited to get an efficient algorithm for solving (4). As long as the problem is not too big, we may assume without loss of generality, that (4) can be solved by any standard nonlinear optimization technique.

To control the degree of convexification and to adjust it with respect to the problem to be solved, Svanberg (1986) introduced so-called moving asymptotes U_i and L_i to replace x_i and $1/x_i$ by

$$\frac{1}{x_i - L_i} \quad , \quad \frac{1}{U_i - x_i}$$

where L_i and U_i are given parameters, which can also be adjusted from one iteration to the next. The algorithm is called method of moving asymptotes. The larger flexibility allows a better convex approximation of the problem and thus a more efficient and robust solution.

References

FLETCHER R., DE LA MAZA E.S. (1987): *Nonlinear programming and nonsmooth optimization by successive linear programming,* Report No. Na/100, University of Dundee, Dept. of Mathematical Sciences, Dundee, U.K.

FLEURY C. (1989): *CONLIN: An efficient dual optimizer based on convex approximation concepts,* Structural Optimization, Vol. 1, 81-89

GILL P.E., MURRAY W., WRIGHT M.H. (1981): *Practical Optimization,* Academic Press

HAN S.-P. (1976): *Superlinearly convergent variable metric algorithms for general nonlinear programming problems,* Mathematical Programming, Vol. 11, 263-282

HAN S.-P. (1977): *A globally convergent method for nonlinear programming,* Journal of Optimization Theory and Applications, Vol. 22, 297-309

HOCK W., SCHITTKOWSKI K. (1981): *Test Examples for Nonlinear Programming Codes,* Lecture Notes in Economics and Mathematical Systems, Vol. 187, Springer

LASDON L., KIM N.-H., ZHANG J. (1983): *An improved successive linear programmin algorithm,* Working Paper 8384-3-1, Dept. of General Business, The University of Austin, Austin, USA

PAPALAMBROS P.Y., WILDE D.J. (1988): *Principles of Optimal Design,* Cambridge University Press

POWELL M.J.D. (1978A): *A fast algorithm for nonlinearly constrained optimization calculations,* in: Numerical Analysis, ed. G.A. Watson, Lecture Notes in Mathematics, Vol. 630, Springer

POWELL M.J.D. (1978B): *The convergence of variable metric methods for nonlinearly constrained optimization calculations,* in: Nonlinear Programming 3, ed. O.L. Mangasarian, R.R. Meyer and S.M. Robinson, Academic Press

SCHITTKOWSKI K. (1980): *Nonlinear Programming Codes,* Lecture Notes in Economics and Mathematical Systems, Vol. 183, Springer

SCHITTKOWSKI K. (1983): *On the convergence of a sequential quadratic programming method with an augmented Lagrangian line search function,* Mathematische Operationsforschung und Statistik, Series Optimization, Vol. 14

SVANBERG K. (1987): *Method of moving asymptotes - a new method for structural optimization,* International Journal on Numerical Methods in Engineering, Vol. 24, 359-373

DESIGN OPTIMIZATION WITH THE FINITE ELEMENT PROGRAM ANSYS^R

by

Günter Müller
Peter Tiefenthaler
Mark Imgrund

Abstract: The paper describes the design optimization procedure implemented in the finite element program ANSYS^R. The strategy is briefly summarized and a number of examples are quoted which show the broad area of application of the ANSYS^R optimization technique.

1. Introduction

Engineering design is an iterative process that strives to obtain a best or optimum design. The optimum design is usually one that meets the design requirements with a minimum expense of certain factors such as cost and weight.

The usual path toward the optimum design is a traditional one. The desired function and performance of the design are first defined. Then, primarily from experience, a trial configuration is developed with the intent of meeting the function and performance requirements. Next, an analysis of the trial arrangement is performed and the results evaluated against the design requirements. The design configuration is then usually altered in an attempt to better meet the design needs. If schedule and budget permit, the cycle begins again.

Design optimization is a mathematical technique that seeks to determine a best design based on criteria set up by the engineer. The technique generates, analyzes, evaluates, and regenerates a series of improving designs until specified criteria are met. The engineer determines the criteria and bounds for the design problem, sets the problem up, but leaves the task of controlling and executing the design cycle to the design optimization routine.

Fully integrated optimization requires software features such as an analysis and optimization database, solid modeling and parametric language. Solid modeling and parametric language are prerequisites for efficient shape optimization so that even complicated 3-D models can be described with a minimum of variables. The ANSYS^R optimization capability encompasses these features. In addition, the optimization module from ANSYS^R can also be used for other purposes such as parametric studies or to curve fit results from analysis or measurements. This method can also be used to correlate experimental results.

This paper presents the design optimization module in the finite element program ANSYS^R. The optimization strategy and the philosophy behind it are briefly described. The numerical methods used in the algorithm are outlined and several examples are cited to show the wide range of this methods' applicability.

2. The Optimization Approach in ANSYS^R

2.1 Philosophy

A basic philosophy of the ANSYS^R optimization capability is that analytical sensitivities should not be required for the optimization method to be implemented. The ANSYS^R approximate optimization (AAO) method discussed in this paper, an alternative to the inefficiencies of the finite differencing method of calculating sensitivities, includes the basic generality of finite differenced sensitivities without incorporating the inefficiencies associated with full reanalysis for each design variable perturbation at each step in the optimization.

The method is applicable to any analysis type in the ANSYS^R program and is completely general since it is not limited to predefined variables with analytical sensitivities. Because the geometric shape, the loads, and the boundary conditions can be parameterized as functions of the design variables, the method is well suited to shape optimization.

The AAO optimization technique presented herein can be described as solving a succession of changing quadratic approximations to the overall design problem while constantly updating these approximations using regression analysis, as steps are taken toward the optimum design. The AAO method borrows approximations and curve fitting techniques which have been known to statistical experimenters for many years.

The rationale behind adopting such a technique is that in many design situations, the chosen objective function and design constraints are more strongly influenced by a handful of design variables that might be chosen to describe the problem. By making use of such dominant variable correlations as early in the optimization strategy as possible, steps may be taken toward an improved design sooner and the influence of the more weakly correlated design variables can be accounted for as additional data points (design vectors) are added to the set.

Since analytical sensitivities are not necessary for the AAO method, it is possible for a wide variety of finite element analysis options and response quantities to be used with this method. Material nonlinearities, such as plasticity, creep, hyperelasticity, viscoelasticity, temperature dependency, etc., geometric nonlinearities such as large rotation, large strain or geometric stiffening, discrete variables, user-defined constraints, and transient or harmonic response are excluded from those quantities which have readily-available analytical sensitivities, but are solvable with the AAO method.

2.2 Definitions

Design Variable - design variables represent those aspects of the design that can be varied to obtain a minimum objective function. Design variables should be independent variables. Design variables are often geometric parameters such as length, thickness, radius, material orientation or even nodal coordinates. Limits on design variables are called side constraints. The full set of design variables is known as the design vector.

State Variable - state variables represent the response of the design to loadings and boundary conditions, and to changes in geometry. Each state variable is a function of one or more of the design variables. Limits placed on state variables are termed behavioral constraints, and act to limit design response and define design feasibility. Stresses, displacements, temperatures, and natural frequencies are typical state variables.

Objective Function - the objective function is a single variable that characterizes the aspect to be minimized. It is a function of one or more of the design variables. Typically weight or cost, it can be virtually any design characteristic desired.

Constrained/Unconstrained Problem - a design problem subject/not subject to limits

Feasible/Infeasible Design - a design that satisfies/violates the constraints (limits)

2.3 The ANSYS[R] Design Optimization Cycle

The optimization module within ANSYS[R] uses approximation techniques to characterize the analysis of a design with a set of quadratic functions at each design loop.

These functions define an approximate subproblem to be minimized, yielding a better design vector for the next design loop. The approximate subproblems are updated at each design loop to account for the additional information, and the process continues until convergence criteria are met. This procedure attempts to gain maximum information from each finite element solution while preserving generality in the choice of design variables, constraints, and objective function.

Following determination of the approximations, the constrained approximate problem is converted into an unconstrained one by using penalty functions. The search for the minimum of the unconstrained problem is then performed using SUMT. As a result of this search, a new design vector is determined. The changes in the objective function and the design variables between this design and the best design yet encountered are evaluated to determine if convergence has occurred and a possible minimum has been reached. Too many sequential infeasible designs or too many design loops (one cycle through design optimization) will cause termination. If neither convergence nor termination occur, the approximations are updated to account for the new design set and the cycle is repeated. This process continues automatically until convergence or termination is indicated. These steps (Fig. 1) are discussed briefly in the following sections. More details can be found in [1].

2.4   The Approximation of the Objective Function and State Variables

By default, the objective function curve in ANSYS^R is approximated in the form of a quadratic equation including cross product terms. The state variables are approximated in the same fashion as the objective function. By using these approximations, highly nonlinear, arbitrary functions can be represented.

Approximation of objective function to be minimized:

$$F = a_{oo} + \sum_{i=1}^{n_d} a_{oi} x_i + \sum_{i=1}^{n_d} a_{ii} x_i^2 + \underbrace{\sum_{i=1}^{n_d-1} \sum_{j=1}^{n_d} a_{ij} x_i x_j}_{i \neq j}$$

Approximation of state variables:

$$g_k = b_{ok} + \sum_{i=1}^{n_d} b_{oik} x_i + \sum_{i=1}^{n_d} b_{iik} x_i^2 + \underbrace{\sum_{i=1}^{n_d-1} \sum_{j=i+1}^{n_d} b_{ijk} x_i x_j}_{i \neq j} \quad (k=1, n_g)$$

Constraints: $\underline{g}_j < g_j < \overline{g}_j \quad (j=1, n_g)$

$\underline{x}_i < x_i < \overline{x}_i \quad (i=1, n_g)$

where

$x_i, x_j \ldots$ *design variable*

$\underline{x} \ldots$ *lower limit;* $\overline{x} \ldots$ *upper limit of design variable*

$n_d \ldots$ *number of design variables*

$\underline{g} \ldots$ *lower limit,* $\overline{g} \ldots$ *upper limit of state variable*

$n_g \ldots$ *number of state variables*

Design optimization with the finite element program ANSYS^R

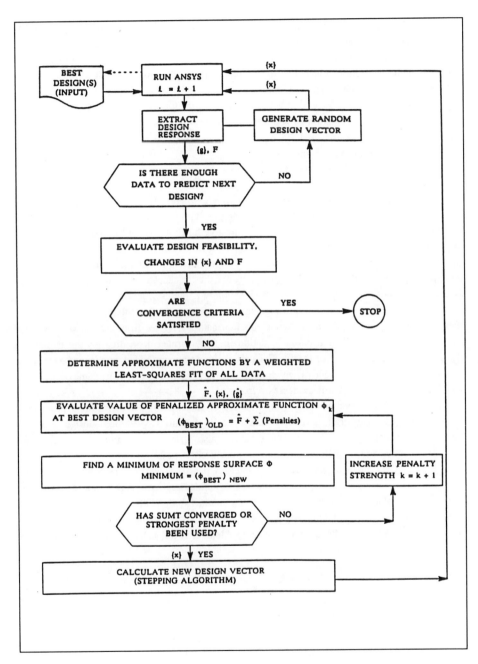

Fig. 1   ANSYS^R Approximate Optimization Method

The coefficients a_{oo}, a_{oi}, a_{ii}, a_{ij} and b_{ok}, b_{oik}, b_{iik}, b_{ijk} are determined by minimizing the weighted least squares error of a set of trial designs.

Since the heart of the AAO method relies on a good approximation to the actual functions in the vicinity of the optimum design, a significant effort is spent on obtaining the best possible equation fit. Weighting factors are used to aid this effort.

The weighting of the design sets can be based on the design variables, the objective function, feasibility, or the product of all three, which is the default relationship in the algorithm. The default weight used for a given design set is

$$W = (W_d \, W_o \, W_f)^n$$

W_d in the expression above is a weight which is inversely proportional to the distance in design space between the design vector and the present best design. W_o is a weight which is unity for the best design and decreases as the objective function value of a design set increases. W_f is a feasibility weighting, which is unity for a feasible design vector, but which decreases inversely proportional to the degree which the design violates constraints. n in this equation is an aging exponent ($1 < n < 5$), designed to reduce the influence of designs encountered early in the optimization process. The full expressions for W_d, W_o, W_f, and n are found in [1].

For each function, a multiple regression coefficient R is calculated. R measures how well the fitted equations match the actual function data and varies from zero to one. At each design loop, either a full or a partial equation fit is performed depending on the number of design sets S and the nature of the fit. For partial fits, terms are added or dropped each loop. Terms are added to the current fit which reduce E. A term is dropped if its removal from the curve fit results in an insignificant decrease in R.

Once an objective function approximation is determined, the constrained problem is changed into an unconstrained one by using penalty functions to enforce design and state variable constraints. The unconstrained function, also termed the response surface, is the objective function approximation function plus penalties.

The AAO method uses extended interior penalty functions in the unconstrained problem. Because this type of penalty function is defined and continuous over all of design space, it is possible to converge to a feasible minimum from outside the feasible region. The objective function approximation and the state variable approximations with their constraints are collectively known as the approximate subproblem. It is this subproblem that is minimized (optimized).

Design optimization with the finite element program ANSYS[R]

2.5 Minimization of the Approximate Problem

The penalized approximate objective function is minimized using the sequential unconstrained minimization technique (SUMT) [2]. This technique searches for the minimum of the current approximation of the objective function.

For each design loop, up to five response surfaces are calculated and minimized, using the result from one response surface as the starting point for the next. The minimum of each response surface is found by a series of line searches in design space similar to the method of feasible directions. See [1] for more information. This sequence continues until the change in the minimum of the response surface is less than a small tolerance. The line searches that occur during this looping process are performed using an iterative algorithm after Powell [15].

Although the numerical methods which are used in the minimization of the approximate subproblem cannot be considered "state of the art", they represent a reliable way to reach the minimum of the quadratic approximation. Since the minimization of the subproblem is such a minor overall contribution to the computational time of the AAO method, little is to be gained by streamlining this step.

2.6 New Trial Design

When the minimum has been determined for the approximate subproblem, a new design is computed. A partial step is taken between the best design encountered so far and the design predicted by minimizing the current approximate subproblem, as follows:

$$\{x\}_{n+1} = \{x\}_n + [a] \{x - x\}_n$$

where: $\{x\}_{n+1}$ = design vector for the next design loop
$\{x\}_n$ = best design vector to date (current loop's design vector)
$\{x\}_n$ = predicted optimal design vector (computed by minimizing the nth approximate subproblem)

In addition, the matrix [a] is a diagonal matrix whose terms are given by:

$$a_i = 1.0 - c_o - c_r R_i$$

In this expression c_o is the previous best design contribution factor, cr is a random contribution fraction and R_i is a random number between -0.5 and +0.5. The factors c_o and c_r are controlled by the user and have defaults of 0.25 for c_o and 0.1 for c_r. The small amount of randomness introduced in this stepping process provides a greater sampling of design space for the equation fits and attempts to prevent ill conditioning of the regression analysis that could happen if the same design point were chosen twice.

2.7 Looping until Convergence or Termination

With the new design variables, another finite element analysis is performed. An improved approximate subproblem is generated which allows a new design vector to be determined. This looping continues until either convergence occurs or the problem terminates because the maximum number of design loops is reached. Convergence is defined to have occurred when all constraints on design variables and state variables are satisfied and the changes in all design variables and the objective function is within a chosen tolerance between loops.

3. Application Exampels

As mentioned earlier, one of the strengths of the AAO method is that it can handle a wide variety of optimization problems. It has been applied to statics, dynamics, acoustics, and heat transfer analysis as well as electrostatics and magnetic field problems. It is even applicable to coupled-field problems, for example thermal-stress or electro-magnetic-thermal stress analysis.

The following examples, touching on the versatility of the method, are taken from various publications of Swanson Analysis Systems, Inc. (the program developer) and from project reports of CAD-FEM GmbH [3, 4, 5, 6, 7, 8, 9, 10, 11, 12, 13, 14]. The nature of each application is described only briefly.

- Minimization of the Stress Intensity in the Fillet of a Housing of a Potentiometer [10]

The goal of this optimization problem was to reduce the maximum stress intensity which occurs in the fillet of a housing. The shape of this fillet was described by four design variables. This optimization problem was constrained only by bounds in the design variables, a state variable was not defined. The maximum equivalent stresses between the initial and the final design were reduced up to 37%. In addition, the stresses along the fillet path were much more homogeneous in the final design (Fig. 2).

Design optimization with the finite element program ANSYS^R

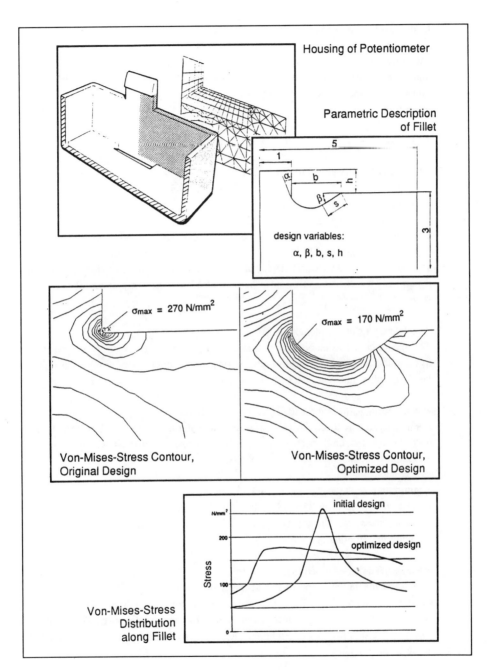

Fig. 2   ANSYS^R Design Opimization

- Optimization of a Connecting Piece under Tensile Load [3]
  The objective function to be minimized was weight. The shape of the model was described by only three design variables, which defined the width in the middle of the structure and the outer contour by using a spline fit. Only one state variable - the maximum allowable stress - was defined.

- Design of a Microstrip Transmission Line [4]
  A microstrip transmission line was to be designed for a known characteristic impedance. The shield and substrate dimensions were fixed. The objective function (and design variable) was the transmission line strip width necessary to produce the desired impedance. The state variable was the characteristic impedance $\pm 1\%$.

- Minimization of the Dimensions of a Coaxial Cable [4]
  The dimensions of a dual dielectric, single conductor coaxial cable was to be minimized. Design variables included the radii of the two dielectric insulators. The state variable was the electric field, not to exceed 40% of the dielectric strength of the materials. The objective function to be minimized was the outer radius of the coaxial cable.

- Axisymmetric Curved Membrane [6]
  The membrane was simply supported at the outside edge and loaded by a concentrated force in the center. The objective was to find the minimum force required to displace the center by 4 mm. The shape of the membrane was defined by cubic splines.

- Minimization of Costs for a Radial Cooling Fin Array [7]
  The objective was to minimize cost as function of fin thickness, fin radius, distance between fins and fluid mean stream velocity. A design constraint was that heat loss was required to be larger than 25000 Watts.

- Minimization of the Weight of a Truck Frame [8]
  Weight was reduced by 25% while limiting the maximum combined stress level and torsional stress level in any beam element. The problem was solved on a personal computer in 510 minutes.

- Frequency Tuning of a Bell [9]
  Given was the initial shape of the Minor Third Bell. The objective was to find the shape for a Major Third Bell. To get the proper tone for a Major Third Bell, the designer had to make sure that given frequency ratios were met. Therefore, the objective function resulted in a minimization of the sum of the squares of the deviation of the frequency ratios.

Design variables were the radii at 9 points, the height and 2 thicknesses of the bell. As state variables the ratio of the frequency of the octave and the frequency of the major 3 were chosen. For this investigation axisymmetric elements allowing nonaxisymmetric loading were chosen. Thus, the problem was reduced by 1 dimension which is an important aspect for reducing the time requirements for optimization runs.

- Failure Optimization of a Thick-walled Pressure Vessel Under Internal Pressure [11]
  For thick pressure vessels, the state of stress or strain is three-dimensional. A 3-dimensional quadratic failure criterion includes interaction among the stress or strain components. In this problem, the Tsai-Wu quadratic interaction failure criterion is used with layered solid elements to optimize the winding angle in a pressure vessel with cylindrical orthotropy (one of the axes of orthotropy is parallel to the longitudinal axis of the cylinder).

- An Acoustic Design Optimization Technique for Automobile Audio Systems [12]
  This example demonstrated the application of an optimization technique to the acoustic design of an audio system. The air inside the cabin enclosure was modeled using the acoustic fluid element of the ANSYS^R program, while the walls were assumed to be rigid surfaces. For a given location of the speakers in the automobile passenger cabin, the orientation of the speakers and the damping of enclosure walls were varied in an attempt to obtain a flat sound level response over a frequency band in the audio range. The objective function of the design was the difference between the maximum and minimum response sound pressure level (SPL) at a given location.

- Design Optimization of Ultrasonic Plastic Welding Equipment [13]
  Both piezoelectric and design optimization capabilities available in the ANSYS^R program were used here to optimize the shape of a booster in ultrasonic plastic welding equipment. The absolute value of the difference between the computed amplitude of the displacement at the bottom of the booster and the expected value was taken as the objective function to be minimized. The first natural frequency of the booster was defined as the only state variable and was bounded between 39.5 and 40.5 KHz. Three independent design variables defined the shape of the structure.

- Fuel Valve Weight Optimization [14]
  This application concerned a cylindrical valve assembly which is used to control the flow of liquid fuel. The electro-magnet assembly consists of a coil, iron core, and armature. When the coil is energized, a magnetic pull force acts to open and close the armature allowing for flow of the fuel. The objective of this optimization was to minimize the overall weight of the valve assembly while maintaining a desired pull force on the armature. The geometry of the assembly was described by five design

variables. The state variable was a minimal acceptable force level. Results of the design optimization showed that a 25% weight reduction was possible.

4. Conclusion

A first step into application of optimization techniques in engineering is described. The technique offered in the ANSYS[R] program combining finite element and optimization technology has been presented. The technique first reduces the global problem to a set of approximated relationships between the objective functions, design variables, and state variables. The reduced problem is then minimized using the sequential unconstrained minimization technique (SUMT). The primary advantage of this approach is that it is not limited to any particular class of problem and that it does not require derivative information.

Although the AAO method implemented in the ANSYS[R] optimization module is very powerful, because every optimization variable can be assigned to any physical meaning, the results may not reflect the absolute optimum or the number of necessary iterations may be quite high. Recognizing that more powerful optimization methods are available, the ability to interface them with the ANSYS[R] program is provided. See [16] for more information on this.

The ANSYS[R] optimization module is implemented in a way that engineers who have no in-depth knowledge of this technique may easily apply it. Though the application is generally still limited to a relatively small number of design parameters due to long computing times, this will be decreasingly important as more powerful computers become readily available.

References

[1]   Kohnke, P.C.: ANSYS[R] Theoretical Manual, Swanson Analysis Systems, Inc., Houston, PA, U.S.A., 1987
[2]   Vanderplaats, G.N.: Numerical Optimization Techniques for Engineering Design, McGraw Hill, New York, 1971
[3]   Tumbrink, H.-J.: Weight Optimization of a Connecting Piece, internal report, Lucas Girling GmbH, Koblenz, 1986
[4]   Ostergaard, D.F.: Adapting Available Finite Element Heat Transfer Programs to Solve 2-D and 3-D Electrostatic Field Problems, IEEE-IAS Society Meeting, Toronto, 1985.
[5]   Beazley, P.K.: ANSYS[R] Design Optimization Seminar Notes, Swanson Analysis Systems, Inc., Houston PA, U.S.A., 1987
[6]   Müller, G., Tiefenthaler, P.: Shape Optimization of an Axisymmetric Membrane, CAD-FEM GmbH internal report, Ebersberg/München, 1988

[7] Imgrund, M.C.: Applying a New Numerical Technique to the Solution of Optimum Convective Surfaces and Associated Thermal Design Problems, Swanson Analysis Systems, Inc., Houston PA, U.S.A., 1985

[8] Johnson, D.H.: Finite Element Optimization of the WABCO 170 Ton Haulpack Truck Frame, Earthmoving Industry Conference, Pretoria, Illinois, U.S.A., 1986

[9] Schweizerhof, K., Müller, G.: Frequency Tuning of a Bell, CAD-FEM GmbH internal report, Ebersberg/München, 1986

[10] Kammerer, R.: Mechanisches Auslegen von Formteilen und Werkzeugen - Wege zu optimalen Ergebnissen mit FEM, Sonderdruck aus "Kunststoffe", Heft 10/1988, Carl Hanser Verlag, München

[11] Marx, F.J., Ambe, P.: ANSYS[R] Revision 4.4 Tutorial Composite Structures, Swanson Analysis Inc., Houston, PA, USA, May 1989

[12] Jamnia, M.A., Rajakumar, C.: An Acoustic Design Optimization Technique for Automobile Audio Systems, SAE Technical Paper Series, Detroit, 1989

[13] Schaller, A.: Design Optimization of Ultrasonic Plastic Welding Equipment with the ANSYS[R] Program, ANSYS[R] 1989 Conference Proceedings, Pittsburgh, PA, May 1989

[14] 30 lb. Solenoid Valve Assembly, Moog, Inc. Space products Division, ANSYS[R] Application Example, Swanson Analysis, Inc., Houston, PA, USA 1988

[15] Powell, M.J.D.: An Efficient Method for Finding the Minimum of a Function of Several Variables Without Calculating Derivatives, Computer Journal, 1964, Vol. 7, pp. 155-162.

[16] Ambe, P., Imgrund, M.: Interfacing the ANSYS[R] Program with an External Design Optimizer, ANSYS[R] TechNotes, Swanson Analysis Systems, Inc. Houston, PA, USA, Sixth Issue, December, 1990.

Günter Müller
Peter Tiefenthaler
CAD-FEM GmbH
Anzinger Straße 11
D-8017 Ebersberg / München

Mark Imgrund
Swanson Analysis Systems, Inc.
P.O. Box 65, Johnson Road
Houston, PA 15342-0065, USA

B&B: A FE–Program for Cost Minimization in Concrete Design

Walter Booz and Georg Thierauf

Abstract. A concept of structural optimization interfacing finite element analysis and mathematical programming is explained for the two stage optimization of concrete structures based on design codes.

1. Introduction

The solution of optimization problems in concrete design using an algorithm for nonlinear programming and a finite element program for static and dynamic analysis of structures (Thierauf 1992), designed primarily for this purpose will be discussed in the following. The optimization is based on German design regulations (DIN 1045, 1988). B&B is available on HP- and IBM-workstations as well as on personal computers with the operating systems MS-DOS or SCO-UNIX. Pre- and postprocessing based on GKS is included.

Actual program limits are:
- 10000 nodes,
- 60000 degrees of freedom,
- 10000 elements,
- 200 cross-section groups.

2. Two stage optimization

2.1 Optimization problems (PG1) and (PL1). In the finite element program B&B every element is assigned to a so-called cross-section group. Elements belonging to such a group all share the property of having the same cross-sectional dimensions and the same reinforcement in concrete design. These dimensions and reinforcement areas constitute the design variables of the optimization problem.

For the design of reinforced concrete structures it is preferable to minimize the total cost (Kirsch 1981) (or an ideal volume) instead of the total weight as in steel structures since self weight is here of major importance.

The objective function of the superior problem (PG1) on structural level can be formulated as follows (Booz 1985):

$$\min \left\{ Z(\underline{y}) = \sum_{i=1}^{nq} \sum_{j \in I_{e_i}} \left[V_{c_j}(\underline{y}) + (k_{1_i} - 1) V_{s_j}(\underline{y}) + k_{2_i} A_{f_j}(\underline{y}) \right] \right\}, \qquad (1)$$

where $\underline{y} \in \mathcal{R}^{nc+ns}$: vector of variables,

$\quad\underline{y}_c \in \mathcal{R}^{nc}$: variable concrete dimensions,

$\quad\underline{y}_s \in \mathcal{R}^{ns}$: variable reinforcement areas,

$\quad nq$: number of cross-section groups Q_i,

$\quad nc = \sum_{i=1}^{nq} m_{c_i}$

$\quad m_{c_i}$: number of variable concrete dimensions of Q_i (Fig. 1),

$\quad ns = \sum_{i=1}^{nq} m_{s_i}$

$\quad m_{s_i}$: number of variable reinforcement areas of Q_i (Fig. 1),

$\quad I_{e_i}$: set of elements belonging to Q_i,

$\quad V_{c_j}$: volume of concrete of element j,

$\quad V_{s_j}$: volume of steel of element j,

$\quad A_{f_j}$: area of formwork of element j,

$\quad k_{1_i}$: ratio of specific cost of steel to the specific cost of concrete of Q_i,

$\quad k_{2_i}$: ratio of the cost of formwork to the specific cost of concrete of Q_i.

For $k_{1_i} = 1$ and $k_{2_i} = 0$, $i = 1, \ldots, nq$ Eq. (1) defines a weight minimization problem. The feasible region is bounded by the following constraints:

- lower and upper bounds of the variables:

$$y_i^- \leq y_i \leq y_i^+, \qquad i = 1, \ldots, nc + ns, \qquad (2)$$

B & B: a FE-program for cost minimization in concrete design

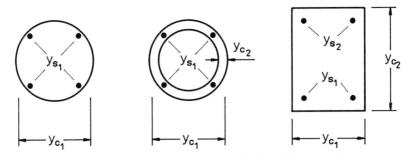

Figure 1: Examples of cross-sections of one-dimensional elements and variables

- constructive or logical constraints (linking of variables):

$$g_{y_j}(\underline{y}) = \sum_k c_k \prod_l y_l + c_j \begin{matrix} \geq 0 \\ = 0 \end{matrix}, \qquad l \in \{1, 2, \ldots, n_c + n_s\}, \qquad (3)$$

where these constraints include the bounds of the reinforcement ratios,

- constraints on the displacements $\underline{r}$:

$$g_{r_j}(\underline{y}) = {}_{perm}r_j - r_j \geq 0, \qquad j = 1, \ldots, nr, \qquad (4)$$

- constraints on the shear stresses $\underline{\tau}$ (w.r.t. German design codes DIN 1045 (1988)):

$$g_{\tau_j}(\underline{y}) = {}_{perm}\tau_j - \tau_j \geq 0, \qquad j = 1, \ldots, nd \cdot nl, \qquad (5)$$

- constraints for failure under bending and axial forces (DIN 1045, 1988):

$$g_{f_j}(\underline{y}) = z_j^* - 1 \geq 0, \qquad j = 1, \ldots, nd \cdot nl, \qquad (6)$$

where nd : number of control points,
nl : number of loading cases.

z_j^* in Eq. (6) are the solutions of an inferior optimization problem (PL1) on cross-section level, i.e.

$$\max \left\{ z(\underline{v}_1) = \frac{1}{f_s(\underline{v}_1)} \frac{M_{\delta_1}^u(\underline{v}_1)}{M_{\bar{y}}^i \cos(\delta_1) - M_{\bar{z}}^i \sin(\delta_1)} \right\}, \qquad (7)$$

where $\underline{v}_1^T = [\varepsilon_{c_1}, \varepsilon_{s_2}, \delta_2]$ are three parameters necessary for a unique description of the linear strain distribution (Fig. 2)

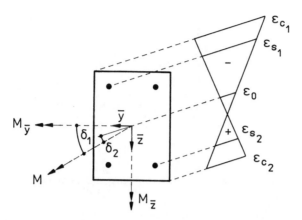

Figure 2: Moments and strains

and ε_{c_1} : minimum concrete strain,

ε_{s_2} : maximum steel strain,

δ_2 : angle between the neutral axis and the inner bending moment,

$\delta_1 = \arctan\left(\dfrac{M_{\bar{z}}^i}{M_{\bar{y}}^i}\right)$,

$M_{\bar{y}}^i, M_{\bar{z}}^i$: inner bending moments,

$M_{\delta_1}^u$: failure moment w.r.t. the inner bending moment's axis (Fig. 2),

$f_s(\underline{v}_1)$: safety factor depending on maximum steel strain (DIN 1045, 1988) with $1.75 \leq f_s \leq 2.1$.

The lower and upper bounds of the strains result from the governing material laws for concrete and for the reinforcement:

$$\varepsilon_{c_1}^- \leq \varepsilon_{c_1} \leq \varepsilon_{c_1}^+ ,$$

$$\varepsilon_{s_2}^- \leq \varepsilon_{s_2} \leq \varepsilon_{s_2}^+ , \tag{8}$$

$$-\pi \leq \delta_2 \leq +\pi .$$

For example, in linear analysis these bounds are prescribed by DIN 1045 (1988) (Fig. 3):

$$\varepsilon_{c_1}^- = -0.0035 ,$$

$$\varepsilon_{s_2}^+ = 0.0050 .$$

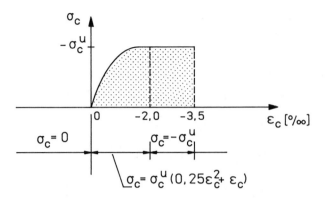
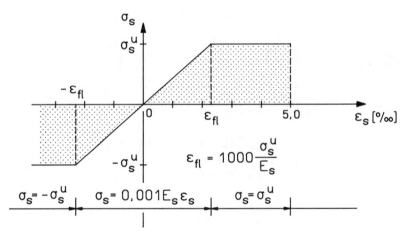

Figure 3: Material laws of concrete and of the reinforcement

For mechanical reasons – a constant stress distribution causes a constant strain – two additional inequality constraints result from the material laws (DIN 1045, 1988), (Fig. 3):

$$g_-(\underline{v}_1) = -\left(1 - \frac{\varepsilon_{c_1}^-}{\varepsilon_{c_g}}\right)\varepsilon_{c_2} + \varepsilon_{c_1} - \varepsilon_{c_1}^- \geq 0 \quad \text{if } N^i < 0,$$
where $\varepsilon_{c_g} = -0.0020$,

$$g_+(\underline{v}_1) = \left(1 - \frac{\varepsilon_{s_2}^+}{\varepsilon_{fl}}\right)\varepsilon_{s_1} - \varepsilon_{s_2} + \varepsilon_{s_2}^+ \geq 0 \quad \text{if } N^i > 0,$$ (9)

where N^i: inner axial force.

These constraints become active at low moments and high axial force level.

Using unique material laws, e.g. for concrete (Dimitrov 1971)

$$\sigma_c = \sigma_c^u \tanh(10^3 \cdot \varepsilon_c), \tag{10}$$

instead of those in DIN 1045 (1988) the inequality constraints (Eqs. (9)) are unnecessary.

In addition, two equilibrium conditions in form of equality constraints must be satisfied:

$$g_N(\underline{v}_1) = z(\underline{v}_1) N^u(\underline{v}_1) - f_s(\underline{v}_1) N^i = 0 ,$$

$$g_{M_\perp}(\underline{v}_1) = M_{\bar{y}}^u \sin(\delta_1) + M_{\bar{z}}^u \cos(\delta_1) = 0 . \tag{11}$$

The forces at failure are defined by:

$$\begin{aligned}
N^u(\underline{v}_1) &= \int_{A_c} \sigma_c(\underline{v}_1) dA + \sum_j A_{s_j} \sigma_{s_j}(\underline{v}_1) , \\
M_{\bar{y}}^u(\underline{v}_1) &= \int_{A_c} \sigma_c(\underline{v}_1) \bar{z} dA + \sum_j A_{s_j} \sigma_{s_j}(\underline{v}_1) \bar{z}_{s_j} , \\
M_{\bar{z}}^u(\underline{v}_1) &= -\int_{A_c} \sigma_c(\underline{v}_1) \bar{y} dA - \sum_j A_{s_j} \sigma_{s_j}(\underline{v}_1) \bar{y}_{s_j} .
\end{aligned} \tag{12}$$

Numerical integration by triangularization of A_c and Gauss quadrature is used for Eq. (12) and therefore it is possible to include any reasonable material law $\sigma(\underline{v}_1)$ and any cross-section.

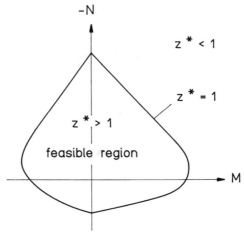

Figure 4: Example of failure surface under bending and axial force

A possible solution of problem (PL1) is shown in Fig. 4.

The relation between the superior optimization problem (PG1) and the inferior optimization problems (PL1) with objective functions and variables is shown in Fig. 5.

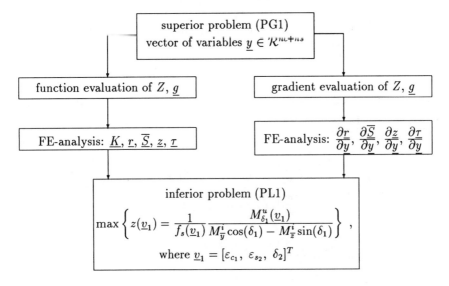

Figure 5: Problems (PG1) and (PL1)

2.2 Optimization problems (PG2) and (PL2). In linear analysis the variables $\underline{y} \in \mathcal{R}^{nc+ns}$ can be separated into the concrete section dimensions $\underline{y}_c \in \mathcal{R}^{nc}$ as independent variables and into the reinforcement variables $\underline{y}_s \in \mathcal{R}^{ns}$ as dependent variables.
Thus the minimization of the steel volume can be reduced to a set of independent inferior problems (PL2) with same hierarchy at cross-section level, where the dimensions of the concrete sections remain constant (Kirsch 1982).
The objective function of this superior problem (PG2) on structural level is that of problem (PG1), see Eq. (1). The constraints are the same, except that the constraints for failure under bending and axial forces (Eqs. (6)) are replaced by constraints on the reinforcement ratios μ:

$$g_{\mu_j} = {}_{perm}\mu_j - \mu_j \geq 0, \qquad j = 1, \ldots, nd \cdot nl, \tag{13}$$

and by the augmented inferior optimization problems (PL2) on cross-section level, i.e.

$$\min \left\{ z(\underline{v}_2) = \sum_{j=1}^{m_{s_j}} A_{s_j} \right\}, \tag{14}$$

with

$$\underline{v}_2^T = [\underline{v}_1^T, \underline{A}_{s_i}^T] \in \mathcal{R}^{3+m_{s_i}}, \tag{15}$$

where $\underline{v}_1$: vector of variables of problem (PL1),
$\underline{A}_{s_i}$: reinforcement of the regarded cross-section.

The bounds for $\underline{v}_1$ are given in Eq. (8), the variable steel areas A_{s_j} are bounded from below by the available areas $A_{s_j}^-$:

$$0 \leq A_{s_j}^- \leq A_{s_j}, \qquad j = 1, \ldots, m_{s_i}. \tag{16}$$

The upper bounds are defined by the permissible reinforcement ratios of the superior problem (PG2) (Eq. (13)). Further constraints for the feasible region are the inequality constraints $g_-(\underline{v}_1)$ and $g_+(\underline{v}_1)$ of problem (PL1) (Eq. (9)) if the material laws of DIN 1045 (1988) are used.

The objective function of problem (PL1) (Eq. (7)) and the two equilibrium conditions (Eq. (11)) are replaced by three equality constraints:

$$\begin{aligned} g_N(\underline{v}_2) &= N^u(\underline{v}_2) - f_s(\underline{v}_1)N^i, \\ g_{M_{\bar{y}}}(\underline{v}_2) &= M_{\bar{y}}^u(\underline{v}_2) - f_s(\underline{v}_1)M_{\bar{y}}^i, \\ g_{M_{\bar{z}}}(\underline{v}_2) &= M_{\bar{z}}^u(\underline{v}_2) - f_s(\underline{v}_1)M_{\bar{z}}^i, \end{aligned} \tag{17}$$

which prevent failure under bending and axial forces of the cross-section, see Eq. (12).

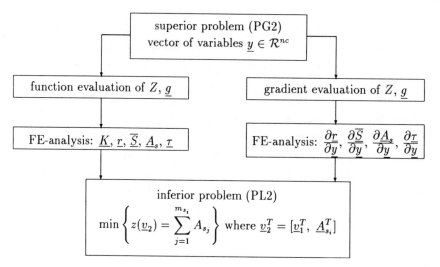

Figure 6: Problems (PG2) and (PL2)

Comparing Fig. 5 and Fig. 6 the differences between problem (PG1) with inferior problems (PL1) and reduced problem (PG2) with augmented inferior problems (PL2) become evident.

3. Linear analysis

The basic equations of the finite element displacement method are:

$$\underline{K}_e(\underline{y})\,\underline{r}(\underline{y}) = \underline{R} + \underline{J}\,\underline{S}_r(\underline{y})\,, \tag{18}$$

where $\underline{K}_e(\underline{y})$: elastic stiffness matrix,
$\underline{r}(\underline{y})$: nodal displacements,
$\underline{R}$: nodal forces,
$\underline{J}$: incidence matrix,
$\underline{S}_r(\underline{y})$: nodal equivalent loads (e.g. self weight).

The stiffness matrix is obtained from (direct stiffness method)

$$\underline{K}_e(\underline{y}) = \underline{J}\,diag\,\{\underline{k}_e(\underline{y})\}\,\underline{J}^T\,, \tag{19}$$

where $diag\,\{\underline{k}_e(\underline{y})\}$ denotes the block diagonal matrix of element stiffness matrices.

The "out of core"-equation-solver used in B&B performs the Cholesky-factorization of the stiffness matrix by submatrices with respect to the skyline of $\underline{K}_e$. At least three submatrices are needed at the same time in the core. The size of these submatrices can also be adjusted. It is therefore possible to solve the system of equations of arbitrary dimensions with this block matrix algorithm on nearly every computer.

With the solution of Eq. (18) and with the incidence matrix $\underline{J}$ the element displacements in global coordinates are obtained:

$$\underline{u}(\underline{y}) = \underline{J}^T \underline{r}(\underline{y})\,. \tag{20}$$

A coordinate transformation of $\underline{u}_i$ results in the displacements of elements i in local coordinates:

$$\underline{\bar{u}}_i(\underline{y}) = \underline{L}_{D_i}^T \underline{u}_i(\underline{y})\,, \tag{21}$$

where $\underline{L}_{D_i}$ is the rotation matrix for element i.

The inner forces $\overline{\underline{S}}_i(\underline{y})$ of element i are computed from the element displacements:

$$\overline{\underline{S}}_i(\underline{y}) = f_i(\underline{y})\underline{E}_i\ \underline{B}_i\ \overline{\underline{u}}_i(\underline{y}) - \overline{\underline{S}}_{R_i}(\underline{y}),\tag{22}$$

where $\underline{E}_i$: material matrix,
$\underline{B}_i$: strain-displacement-matrix,
$f_i(\underline{y})$: cross-section area or moment of inertia.

With the inner forces the inferior problems, either (PL1) or (PL2) (Chap. 2) can be solved. Subsequently the constraints for failure (Eq. (6)) in the superior problem (PG1) or the objective function (Eq. (1)) and the constraints on the reinforcement ratios (Eq. (13)) in the superior problem (PG2) can be calculated.

4. Sensitivity analysis

Partial differentiation of Eq. (18) with respect to $\underline{y}$ results in the linear equation system

$$\underline{K}_e\frac{\partial}{\partial \underline{y}}\underline{r} = \frac{\partial}{\partial \underline{y}}\underline{R} - \underline{J}\ diag\left\{\frac{\partial}{\partial \underline{y}}\underline{k}_e\right\}\underline{J}^T\underline{r} + \underline{J}\frac{\partial}{\partial \underline{y}}\underline{S}_R,\tag{23}$$

which can be solved with the factorized $\underline{K}_e(\underline{y})$ by forward-backward-substitution. Using this direct method one can derive analytical gradients for the displacements if the right hand side of Eq. (23) is determinated analytically w.r.t. $\underline{y}$:

$$\frac{\partial}{\partial \underline{y}}\underline{r} = \underline{K}_e^{-1}\left[\frac{\partial}{\partial \underline{y}} - \underline{J}\ diag\left\{\frac{\partial}{\partial \underline{y}}\underline{k}_e\right\}\underline{J}^T\underline{r} + \underline{J}\frac{\partial}{\partial \underline{y}}\underline{S}_R\right].\tag{24}$$

Gradients for the inner forces can also be given:

$$\frac{\partial \overline{\underline{S}}_i}{\partial \underline{y}} = \underline{E}\ \underline{B}\left(\frac{\partial f_i}{\partial \underline{y}}\overline{\underline{u}}_i + f_i\frac{\partial \overline{\underline{u}}_i}{\partial \underline{y}}\right) - \frac{\partial \overline{\underline{S}}_{R_i}}{\partial \underline{y}}.\tag{25}$$

The gradients w.r.t. $\underline{y}$ of the iterative solutions of the optimization problems (PL1) or (PL2) on cross-section level must be calculated semi-analytically. This is done by a second-order central-difference approximation. For that purpose one needs the vectors $\overline{\underline{S}}_i(\underline{y} + \Delta\underline{y})$ and $\overline{\underline{S}}_i(\underline{y} - \Delta\underline{y})$.
A Taylor-series expansion of $\overline{\underline{S}}_i(\underline{y} \pm \Delta\underline{y})$ up to the linear term yields

$$\overline{\underline{S}}_i(y_j \pm \Delta y_j) \cong \overline{\underline{S}}_i(y_j) \pm \Delta y_j \frac{\partial \overline{\underline{S}}_i}{\partial y_j}(y_j)\tag{26}$$

$$= \underline{E}\ \underline{B}\left(f_i\overline{\underline{u}}_i \pm \Delta y_j\frac{\partial f_i}{\partial y_j}\overline{\underline{u}}_i \pm \Delta y_j f_i\frac{\partial \overline{\underline{u}}_i}{\partial y_j}\right) - \underline{S}_{R_i}(y) \mp \Delta y_j\frac{\partial \overline{\underline{S}}_{R_i}}{\partial y_j}(y_j)$$

for all elements i and all variables j.

With $\overline{S}_i(\underline{y} \pm \Delta \underline{y})$ the inferior problems, either (PL1) or (PL2) are solved. Subsequently the gradients of the constraints for failure (Eq. (6)) in the superior problem (PG1) or the gradient of the objective function (Eq. (1)) and the gradients on the reinforcement ratio constraints (Eq. (13)) in the superior problem (PG2) can be calculated.

5. Solution of the optimization problems

5.1 Optimization algorithm. In the solution of the problems discussed before the following requirements should be fulfilled:

- global convergence, in order to arrive always at a permissible solution,
- as few function evaluations as possible, since every function evaluation requires a complete FE-analysis (Figs. 5 and 6),
- robustness with respect to numerical approximations involved in the calculation of gradients (Chap. 4).

A SQP algorithm satisfies all these requirements (Schittkowski 1982a).

The optimization package NLPQL (Schittkowski 1991) used in B&B possesses the additional advantage that the function and gradient evaluations can be called outside of the NLPQL-subroutines (Fig. 7).

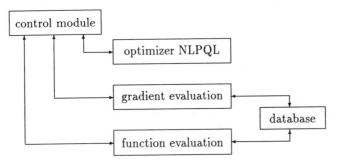

Figure 7: Flow-chart, moduls

This modular structure is an important feature of the two stage optimization of reinforced concrete (Chap. 2), as all moduls share the same NLPQL-subroutines.

5.2 Scaling.
In order to get all terms of the optimization problems in Chap. 2 free of dimensions, variables and functions of the superior problems (PG1) and (PG2) are scaled.

The objective function is scaled by the starting value $Z(\underline{y}^\circ)$:

$$\hat{Z}(\underline{y}^j) = \frac{Z(\underline{y}^j)}{Z(\underline{y}^\circ)} . \tag{27}$$

The variables (Eq. (2)) are scaled by (Lawo 1987):

$$0 \leq \hat{y}_i = \frac{y_i - y_i^-}{y_i^+ - y_i^-} \leq 1 , \qquad i = 1, \ldots, n , \tag{28}$$

and the constraints on the displacements (Eq. (4)), the constraints on the shear stresses (Eq. (5)) and the constraints on the reinforcement ratios (Eq. (13)) are scaled by the permissible values:

$$\hat{g}_{r_j}(\underline{y}) = 1 - \frac{r_j}{perm\, r_j} \geq 0 , \qquad j = 1, \ldots, nr , \tag{29}$$

$$\hat{g}_{\tau_j}(\underline{y}) = 1 - \frac{\tau_j}{perm\, \tau_j} \geq 0 , \qquad j = 1, \ldots, nd \cdot nl , \tag{30}$$

$$\hat{g}_{\mu_j}(\underline{y}) = 1 - \frac{\mu_j}{perm\, \mu_j} \geq 0 , \qquad j = 1, \ldots, nd \cdot nl , \tag{31}$$

Eqs. (11) and (22) of the inferior problems the constraints g_N are scaled by a factor $A_c \cdot \sigma_c^u$ and the constraints g_M by a factor $A_c^{1.5} \cdot \sigma_c^u$.

5.3 Numerical gradients.
Δy_j in Eq. (26) is given by the user and should be

$$10^{-5} \leq \Delta \hat{y}_j \leq 10^{-3} \tag{32}$$

related to the scaled variables (Eq. (28)).

In the inferior problems (PL1) and (PL2) on cross-section level, all gradients are calculated by a first-order forward-difference approximation with a fixed value of

$$\Delta v = 10^{-4} . \tag{33}$$

5.4 Convergence criteria.
The stopping criteria of NLPQL is described by Schittkowski (1982b).

Depending on the values for $\Delta \hat{y}_j$ (Eq. (32)), a suitable interval for ε_G in the superior problems (PG1) and (PG2) is

$$10^{-5} \leq \varepsilon_G \leq 10^{-3} . \tag{34}$$

In the inferior problems (PL1) and (PL2) ε_L is fixed to

$$\varepsilon_L = 10^{-4} . \tag{35}$$

6. Example

The example is an angular slab with eccentrically connected stiffening beams (Fig. 9) where only a linear analysis is carried out to get the inner forces. It is similiar to an example by Thierauf (1989).

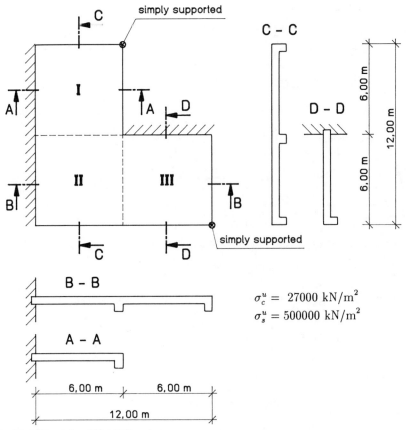

Figure 9: Angular slab with stiffeners

The FE-discretization by 481 nodes with a total of 2223 degrees of freedom and by 432 DKQ-elements (Batoz 1982) with 84 stiffening beam elements (Thierauf 1992) as well as the subdivision in cross-section groups (= design groups) is shown in Fig. 10. This subdivision yields the following optimization variables:

- shell elements:

$$y_c^i, \; y_{s_j}^i, \qquad i = 1, \ldots, 12, \quad j = 1, \ldots, 4,$$

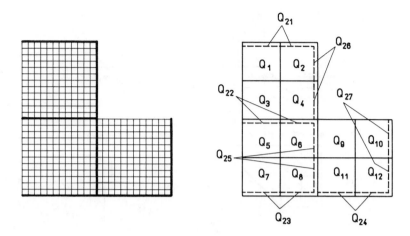

Figure 10: FE-discretization and cross-section groups

- beam elements:

$$y_{c_1}^i, \quad y_{c_2}^i, \quad y_{s_1}^i, \quad y_{s_2}^i, \qquad i = 21, \ldots, 27.$$

Thus,

$$\underline{y}_c \in \mathcal{R}^{26}, \quad \underline{y}_s \in \mathcal{R}^{62} \text{ and } \underline{y} \in \mathcal{R}^{88},$$

and by combining the following loads, 8 loading cases are obtained:

- self weight $g_i = \begin{cases} 24 \cdot y_c^i & \text{kN/m}^2 \quad i = 1, \ldots, 12, \\ 24 \cdot y_{c_1}^i \cdot y_{c_2}^i & \text{kN/m} \quad i = 21, \ldots, 27, \end{cases}$

- dead load $\widetilde{g} = 2$ kN/m^2,

- live load $p = 5$ kN/m^2 at different positions.

That means, in contrast to Thierauf (1989), loading depends on the optimization variables.

Due to the eccentrical connections, the stiffness matrices of the beam elements depend not only on their cross-sections but also on the variable thickness of the corresponding shell elements (Fig. 11).

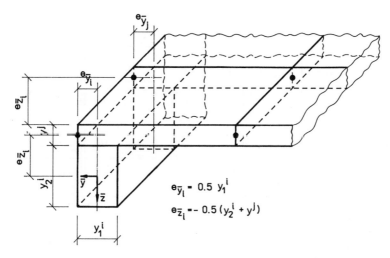

Figure 11: Plane shell element and eccentrically connected beam element with rectangular cross-section

The following constraints are considered:

- lower and upper bounds of the variables:

    $$0.15 \text{ m} \leq y_c^i \leq 0.40 \text{ m}, \qquad i = 1, \ldots, 12, 21, \ldots, 27,$$

- linking of variables:

    $$\begin{aligned}
    y_c^1 &= y_c^2 = y_c^3 = y_c^4, \\
    y_c^5 &= y_c^6 = y_c^7 = y_c^8, \\
    y_c^9 &= y_c^{10} = y_c^{11} = y_c^{12}, \\
    y_{c_j}^{21} &= y_{c_j}^{22} = y_{c_j}^{23} = y_{c_j}^{24}, \\
    y_{c_j}^{25} &= y_{c_j}^{26} = y_{c_j}^{27},
    \end{aligned} \qquad j = 1, 2,$$

- constraints on the shear stresses:

    the permissible shear stresses are given in DIN 1045 (1988).

- constraints on the reinforcement ratios:

    $\mu \leq 0.02$ for the shell elements,
    $\mu \leq 0.04$ for the beam elements.

One control point for each element results in $(432 + 84) \cdot 8 = 4128$ constraints on the shear stresses as well as on the reinforcement ratios. Therefore the number of these constraints is reduced by searching for the minimum of the constraints for all loading cases and for all control points of each cross-section group, although here the constraints might become non-differentiable. Thus one gets 19 constraints on the shear stresses as well as on the reinforcement ratios.

A step-width of

$$\Delta \hat{y}_j = 10^{-3}, \qquad j = 1, \ldots, 26,$$

is choosen for the gradients (Eqs. (26), (32)) and for the stopping criteria (Eq. (34))

$$\varepsilon_G = 10^{-3}$$

is assumed.

The results for different cost ratios k_1 is shown in Tab. 1. Because of its small influence the cost of formwork is neglected.

k_1	1	10	30	50	70	90
V [m³]	18.20	18.40	19.43	20.75	22.99	25.82
Z [m³]	18.20	21.76	29.23	35.29	41.26	46.94
y_c^1 [m]	.1500	.1500	.1568	.1629	.1951	.2194
y_c^5 [m]	.1500	.1500	.1500	.1618	.1778	.2182
y_c^7 [m]	.1500	.1500	.1649	.1673	.1756	.1850
$y_{c_1}^{21}$ [m]	.1500	.1500	.1557	.2041	.2145	.2252
$y_{c_2}^{21}$ [m]	.4000	.4000	.4000	.4000	.4000	.4000
$y_{c_1}^{25}$ [m]	.1500	.1500	.1500	.1500	.1641	.1735
$y_{c_2}^{25}$ [m]	.2087	.2812	.3529	.4000	.4000	.4000

Table 1: Results of the optimization

The optimization problem was solved with several different starting points. All yield the same solutions which seem to be global optima. The average number of iterations was 8.

7. Conclusion

For the optimization of concrete structures it is usually necessary, because of the dimensions of the problems involved, to apply the seperation of variables into dependent and independent ones as discussed here.
The results show, that for reinforced concrete structures the difference between weight- and cost-minimization is considerably.

References

Thierauf G. (ed). (1992), 'B&B - A FE-Program for Analysis and Automatic Design of Structures - Theoretical manual and user's guide', Forschungsberichte aus dem Fachgebiet Bauwesen, Universität GHS Essen.
DIN 1045 (7/1988), Beton und Stahlbeton - Bemessung und Ausführung, Beuth-Verlag, Berlin.
Kirsch U. (1981), Optimum Structural Design, McGraw Hill, New York.
Booz W. (1985), Zweistufenoptimierung von Stahlbetontragwerken mit Hilfe der sequentiellen quadratischen Programmierung, Dissertation, Universität GHS Essen.
Dimitrov N., Herberg W. (1971), Festigkeitslehre I, Walter de Gruyter, Berlin/New York.
Kirsch U. (1982), 'Multilevel optimal design of reinforced concrete structures', in Eschenhauer, H. and Olhoff, N. (eds.), Optimization methods in structural design, Euromech-Colloquium 164, University of Siegen, FR Germany, pp. 157–161.
Schittkowski K. (1982a), 'Theory, implementation, and test of a nonlinear programming algorithm', in Eschenhauer, H. and Olhoff, N. (eds.), Optimization methods in structural design, Euromech-Colloquium 164, University of Siegen, FR Germany, pp. 122–132.
Schittkowski K. (1991), EMP: An Expert System for Mathematical Programming, Mathematisches Institut, Universität Bayreuth.
Lawo M. (1987), Optimierung im Konstruktiven Ingenieurbau, Vieweg-Verlag, Braunschweig/Wiesbaden.
Schittkowski K. (1982b), On the Convergence of a Sequential Quadratic Programming Method with an Augmented Lagrangian Line Search Function, Technical Report SUL 82-4, Dept. of Operations Research, Stanford University, Standford, USA.

Thierauf G. (1989), Optimale Bemessung und Optimierung im Konstruktiven Ingenieurbau, Bauingenieur 64, pp. 463–472.

Batoz J.-L., Tahar M. B. (1982), Evaluation of a new quadrilateral thin plate bending element, International Journal for Numerical Methods in Engineering, vol. 18, pp. 1655–1677.

Author's address:

Dr.-Ing. Walter Booz
Prof. Dr.-Ing. Georg Thierauf
FB 10 – Baumechanik/Statik
Universität – GHS – Essen
Universitätsstr. 15
D-4300 Essen 1
Telefon (0201) 183-2672
Telefax (0201) 183-2675

The CAOS System

John Rasmussen, Erik Lund, Torben Birker & Niels Olhoff

Abstract. The facilities of the structural optimization system CAOS are presented. CAOS is a system for CAD-integrated structural optimization and rational design developed at Aalborg University. The geometrical definition is based on the Auto-CAD® system, and the optimized geometry can be transferred back into the Auto-CAD model and replace the original component.

1. Introduction

When the development of the Computer Aided Optimization System (CAOS) was initiated in 1986, the aim was merely to create a framework for experimentation with various techniques of structural shape optimization, i.e., to extend the large amount of theoretical research in this field (see Olhoff and Taylor 1983) into practical, computer oriented applications.

Around this time, several papers on optimization *systems* (rather than just algorithms) capable of solving a larger variety of problems, were beginning to emerge. Some of the foremost examples are Esping (1984), Braibant and Fleury (1984), Bennet and Botkin (1985), Botkin et. al. (1986) and Stanton (1986). Extensive overviews of available systems have been produced by Ding (1986) and Hörnlein (1987).

These papers among others gave birth to the idea of improving the practical applicability of CAOS by interfacing it with a CAD system, and taking advantage of the CAD system's interactive graphical facilities in defining the problem. This idea was in concordance with the commercial tendency towards integrated CAD/CAE systems. It was decided to investigate the possibilities for extending these ideas into the field of structural optimization.

Since then, CAOS has evolved into a full three-dimensional CAD-integrated structural optimization system capable of solving complex problems of shape optimization and, by the use of a CAD system, to provide the user with opportunity to control, modify, visualize, and, above all, manufacture the result of the optimization process.

The widely used commercial AutoCAD® system is used as the basis for CAOS, but the system concept is independent of the AutoCAD data structure and the techniques used in CAOS can therefore be applied in connection with most other CAD systems as well.

1.1 General outline. CAOS has been under constant development over a period of four years and is today a fully operational shape optimization system with a number of interesting features.

1. One objective function and any number of constraints may be picked among the following functions:

    Volume
    Weight
    Mass moment of inertia
    Area moment of inertia
    Elastic displacement of selected material points
    Maximum elastic displacement of any point
    Compliance
    Stress (several types) in selected material points
    Maximum stress (several types) at any point

2. The analysis facilities producing the above mentioned function values incorporate

    Plane stress/strain elements
    Axi-symmetric elements
    Solid brick elements
    Static linear-elastic analysis
    Static heat transfer analysis
    Thermo-elastic analysis

3. CAOS has been interfaced with the topology optimization system HOMOPT by Bendsøe (1989), thus enabling any problem to be pre-processed in order to find a good starting topology for the shape optimization problem. An example of this integration will be presented in section 6.2.

CAOS is thus a rather general system for structural shape optimization. However, the notable difference between CAOS and other systems for structural optimization is its full integration with the AutoCAD system. The definition of an optimization problem including the finite element model, application of loads and boundary conditions, and specification of objective function and constraints, takes place entirely inside the CAD system utilizing the highly developed interactive graphical facilities available there, and the optimized geometry can be transferred back into the CAD model and take the place of the original geometry. This is a major advantage as shape optimized structures often have complicated shapes which are difficult to describe and transfer manually. Also from the fabricational point of view, the CAD integration is preferable because it enables the manufacturing of the often complicated optimized geometries by numerically controlled machines programmed via a CAM (Computer Aided Manufacturing) interface.

1.2 Software organization. CAOS is based on the important distinction between *design model* and *analysis model*. The design model is a variable description of the shape of the structure. It is closely connected with the CAD model and totally distinct from the finite element model that is used for the analysis. In the case of CAOS, the design model consists of so-called design elements which topologically are plane quadrangles, or three-dimensional hexahedrons, which together make up the geometry.

All definitions, including boundary conditions and other analysis-related information, specified by the user refer to the design model. The user adds specifications to the design model describing the desired number of finite elements and eventual concentrations of nodes in critical regions, but the link between loads, boundary conditions and the actual finite element model is established automatically by CAOS. This provides the attractive feature that it is very easy to change the finite element model if the user suspects that it is inadequate. It is simply a matter of changing the specification of the mesh. Loads and boundary conditions are automatically reorganized to reflect the new analysis model. This system effectively enables the user to run the first part of the optimization on a comparatively crude model which can be refined toward the final stages of the optimization process.

Sensitivity analysis and subsequent improvement of the geometry is performed on the design model. This means that the optimization process is less susceptible to try improving the result by corrupting the analysis model as is often the case when locations of finite element nodes are used directly as geometrical parameters of the model, see, e.g., Rodrigues (1988).

The distinction between the design and analysis models is reflected in the organization of the software and the data which it processes. The system consists of a number of programs which, in the case of a purely elastic problem, without heat conduction, interact via binary and ASCII files in the following way:

1. The result of the CAD definition of the problem is automatically saved on a number of simple text files. Each time a CAOS function has been executed in the CAD system, the corresponding text file, and thereby the optimization model, is updated.

2. A preprocessor reads the information from these files and performs a number of operations which need be performed only once for each problem and mesh topology:
    - Various checks of the validity of the problem.
    - Formation of equation numbers and imposition of boundary conditions.
    - Assignments between the parameters of the model and the design variables of the problem.

3. A mesh generator reads the current values of the design variables, updates the geometry and generates the current analysis model and force vector (which may vary with the design).

4. A finite element module performs a traditional displacement-based analysis of the system and evaluates nodal displacements and stresses.

5. For each design variable, the preprocessor perturbs the geometry and the finite element module evaluates displacement and stress sensitivities. The sensitivity analysis will be further discussed in section 4.

6. A sorting algorithm extracts the necessary results according to the specifications of objective function and constraints and generates an input tableau for the optimizer. The formation of this tableau will be explained in section 2.2.

7. A standard SIMPLEX algorithm is used to solve the subproblem and generate an improved set of design variables.

8. If the problem has not converged, go to 3.

1.3 Post processing. The PC-version of CAOS has a built-in menu driven post processor which has the following functions, among others:

- Visualization of the three-dimensional finite element model from any view point
- Removal of selected elements for investigation of the interior of the model
- Visualizations with hidden lines removed
- Display of node and element numbers
- Deformed geometry visualizations
- Color and level curve representation of stress and stress sensitivity fields
- Generation of HPGL plot files for hardcopy generation or export to desktop publishing systems

The post processor works with EGA, VGA and Hercules graphics adapters. In the case of monochrome Hercules graphics, color representations of stress fields are replaced by hatching patterns on the surface of the structure. The post processor is self explanatory, extremely easy to use, and can be operated with a mouse. It may be invoked at any time in the optimization process to display the current geometry and related analysis results.

1.4 Availability. The CAOS system has previously exclusively been an academic, non-commercial system. However, starting 1 January 1992, the system will be commercialized through *Rational Engineering, Peder Barkesgade 40 st., DK-9000 Aalborg, Denmark. Tel. + 45 98 11 68 47*. For further information, please contact this company or the authors.

The CAOS system

Installations will are readily available for personal computers running the DOS operative system and equipped with mathematical co-processor and harddisk. In this environment, the system can handle up to 1400 nodes in two dimensions and up to 1000 nodes in three dimensions. Workstation and mainframe installations without these limitations are individually negotiable. Large discounts are given to research and educational institutions.

In order to take advantage of the CAD integration, AutoCAD must be installed on the computer. It is possible to perform the definition of the problem using AutoCAD on a PC, transfer the data to a workstation or mainframe for the actual optimization, and postprocess results on a PC. The topology optimization module HOMOPT is not commercially available, but a dedicated topology optimization module for CAOS will be available in forthcoming versions of the system.

2. Optimization model

CAOS is geometrically a very general system. It is in principle possible to handle any two- or three dimensional problem, but, as with any other system, the problem of creating a good optimization model grows considerably with the complexity of the initial geometry. CAOS is a foremost a shape optimization system, but sizing is also covered in the form of thicknesses in 2D models. Material property optimization is not included at present.

2.1 Design variables. The geometrical description in CAOS is based on design elements which may be two- or three-dimensional. A two-dimensional example is shown in figure 1. In the 2D case, a design element consists of four curves forming a quadrangular geometry. In the 3D case, each design element is built from 12 curves which form a hexahedron. Except from the additional z-coordinate, there is no difference in defining 2D and 3D curves in CAOS.

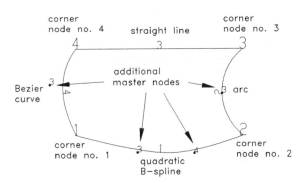

Figure 1. Quadrangular design element with four different boundary shape forms.

Each curve is selected by the user from a library of curve types:

1. Piecewise straight lines
2. Circular arcs

3. Ferguson splines (natural cubic splines)
4. B-splines
5. Bezier curves

In picking the curve types, the user can control the possible outcome of the optimization. For instance, holes in the structure are often required to be circular for fabricational reasons. When built from circular arcs, such holes can participate in the shape optimization and change size and position but will remain circular.

The shape of a design element is given by the shape of its boundaries. The shapes of the boundaries are controlled by the use of so-called master nodes as indicated in figure 1, and the shape of the structure can be controlled through the positions of the master nodes.

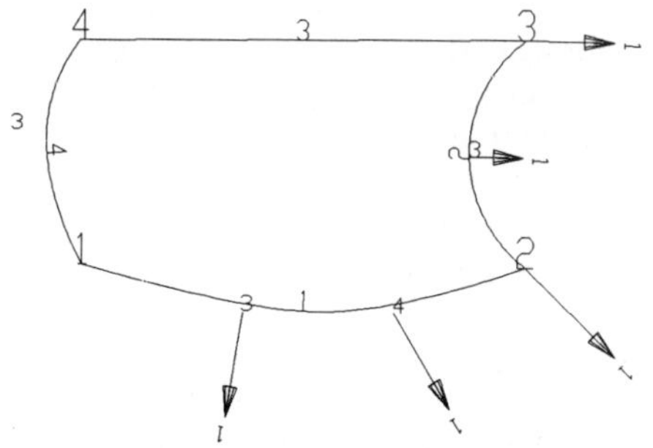

However, it turns out that the spatial coordinates of the master nodes are not directly a good choice for design variables because they do not provide the user with much flexibility in defining the design space, and because they often lead to the use of more design variables than necessary. CAOS overcomes this problem by the use of so-called move directions (figure 2). These are translational operators assigned to the master modes. One or more move directions with corresponding design variables can be assigned to each master node. The master node can only move along or opposite its move directions at a maximum distance specified by the upper and lower values of the design variables.

Figure 2. Variable model obtained by giving master nodes move directions.

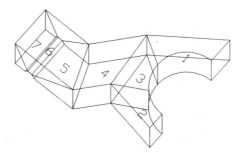

Figure 3. 3D model of one quarter of bearing pedestal modelled by 7 design elements.

Move directions may be linked such that they are assigned the same design variable. This mechanism is useful for imposing geometrical symmetry on the solution.

All definitions of design elements, boundaries, move directions, and design variables take place inside the CAD system giving the user the benefit of the interactive graphical capabilities, i.e., the pictures of design elements in figure 1 – 3 are hardcopies of the AutoCAD screen when working with CAOS. Figure 3 shows how several design elements may be combined to form complex geometries.

2.2 Mathematical formulation.

The mathematical formulation of the shape optimization problem is as follows:

Minimize

$$g_0(x_j) \qquad (j = 1..n-1)$$

Subject to (1)

$$g_i(x_j) \leq G_i \qquad (i = 1..m, \; j = 1..n-1)$$

$$x_j^{min} \leq x_j \leq x_j^{max} \qquad (j = 1..n-1)$$

where $n-1$ is the number of so-called design variables, x_j, and m is the number of constraints. The objective and constraint functions g_i ($i = 0..m$) are specified by the user as a part of the problem formulation. They can, as previously mentioned, be picked and combined freely from a library with the following contents:

1. Volume
2. Weight
3. Mass moment of inertia
4. Area moment of inertia
5. Elastic displacement of selected material points
6. Maximum elastic displacement of any point
7. Compliance
8. Stress (several types) in selected material points
9. Maximum stress (several types) in any points

Mathematically, different entries in this list lead to very different optimization problems. Entries 5 and 8 are ordinary scalar functions which can be derived directly from the output of the finite element analysis. Entries 1 through 4 and 7 are of integral type and require some postprocessing of the results to be evaluated. Entries 6 and 9 lead to min/max-problems with non-differentiable objective functions.

However, CAOS makes use of the so-called bound formulation presented by Olhoff (1989). This formulation enables the system to handle the optimization problem in a uniform way regardless of the blend of scalar-, integral- and min/max-criteria defined by the user.

Given the min/max objective function $g_0 = \max(g_k)$, $k=1..p_0$, and a number of constraints, $g_i = \max(g_{ik}) \leq G_i$, $i = 1..m$, $k = 1..p_i$, we get the following bound formulation of the problem:

Minimize   β
x_j, β

Subject to (2)

$$w_i g_{ik}(x_j) - \beta \leq 0 \qquad (i = 0..m, \; k = 1..p_i, \; j = 1..n-1)$$

$$x_j^{min} \leq x_j \leq x_j^{max} \qquad (j = 1..n-1)$$

An extra design variable, β, has been introduced, rendering the total number of variables to n. By m we designate the original number of constraints regardless of whether these are scalar-, integral- or min/max-functions. The number of points whereby a min/max-condition i is represented, for instance the number of nodal stresses among which the maximum stress is to be found, is termed p_i. This number is obviously 1 if condition i is scalar or integral. The weight factors w_i are imposed on the constraints to allow them to be limited by the same β-value as f. Prior to the call of the optimizer, w_i is found from the relation:

$$G_i w_i = \beta_0 \implies w_i = \frac{\beta_0}{G_i} \qquad (3)$$

where β_0 is the present value of β.

Tableau (2) is valid regardless of the blend of functions g_i, i = 0..m, and the numerical operations performed are therefore identical for any problem that the user could possibly define. The bound formulation has greatly simplified the programming of the optimization module of CAOS.

The problem (2) is non-linear and implicit. This means that we can find values of the objective and constraint functions for given values of the design variables, x_j, but we do not know the explicit expressions for these functions. In order to solve (2), CAOS, like other structural optimization systems, works with successive explicit approximations of the problem. In CAOS, a sequential linear programming (SLP) approach is used, and the corresponding linear problem is solved by a SIMPLEX algorithm. Previously, the Method of Moving Asymptotes (MMA) by Svanberg (1987), was also included in the system. MMA is a dual method which is very well suited for the combination of "many" design variables and "few" constraints. However, in the presence of stress constraints, the opposite is usually the case, and in connection with CAOS, SLP has been found more reliable.

The SLP input is obtained by approximating (2) with linear Taylor expansions originated at the current point, x_0, with respect to the actual design variables, x_j and the artificial bound variable, β. Performing this linearization and collecting the constant terms on the right hand side of the inequalities, we get the problem:

Minimize $\quad\beta$
x_j, β

Subject to

$$w_i \sum_{j=1}^{n-1} \frac{\partial g_{ik}}{\partial x_j} x_j - \beta \leq w_i \left[\sum_{j=1}^{n-1} \frac{\partial g_{ik}}{\partial x_j} x_{0j} - g_{ik}(x_0) \right], \quad (i = 1..m, \ k = 1..p_i) \tag{4}$$

$$x_j^{\min} \leq x_j \leq x_j^{\max}, \quad (j = 1..n-1)$$

Which, provided the derivatives are known, is explicit and can be solved by a suitable SIMPLEX algorithm. For a discussion on the evaluation of derivatives please refer to section 4.

3. Analysis

The CAOS finite element code is equipped with the following element types:

1. Plane stress, 4 node bilinear
2. Plane strain, 4 node bilinear
3. Axi-symmetric, 4 node bilinear
4. 3D brick element, 8 node.

These elements may be used for static analysis of thermal, elastic and thermo-elastic problems, i.e., the temperature stresses may be calculated, superimposed on the geometry, and take part in an optimization problem in the sense that changes in the stress and displacement field due to changes in the temperature field are included in the formulation of the optimization problem.

3.1 Boundary conditions. In the case of a thermal problem, the user may specify

1. Temperature
2. Convection coefficient + environment temperature
3. Flux

at any point or over any part of the boundary. In the case of a distributed condition, local values are interpolated between two end point values in the 2D case and four corner values in the 3D case.

In the case of an elastic problem, the following condition types apply:

1. Specified displacement of a point or an interval of the boundary
2. Zero displacement (roller skate condition) in one or more coordinate directions.
3. Concentrated and distributed forces.
4. Centrifugal forces
5. Gravitational forces.

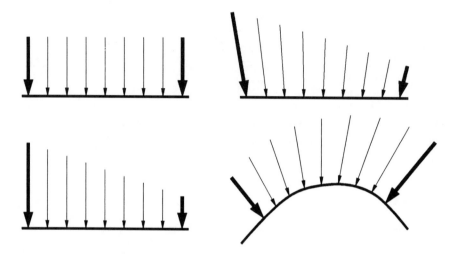

Figure 4. A number of possible load distributions. The arrows in bold are the ones actually defined in CAOS.

For distributed conditions, intermediate values are interpolated between specified end point values, as shown in figure 4, such that complicated force distributions may be conveniently defined, eventually by combinations of several specifications.

3.2 Analysis results. From the analysis point of view, CAOS is a simple system. Only linear static analysis is included, and materials must be isotropic. It is possible to define different materials in various parts of the geometry and to optimize the shape of interfaces between them.

As stress criteria, depending on the coordinates of the problem, the following types are included:

1. Coordinate directions, i.e., σ_{xx}, σ_{yy}, σ_{zz}, σ_{rr}, $\sigma_{\theta\theta}$
2. Maximum shear stress
3. Principal stresses
4. von Mises reference stress

These stress types may be used as objective functions and/or constraints, and they may all be visualized by the post-processor. Furthermore, temperatures may be visualized, and stresses arising from the temperature distribution may be included in the optimization problem.

The system does not presently handle dynamic problems or stability.

4. Sensitivity analysis

The optimization in CAOS is performed in an iterative [Analysis]–[Sensitivity Analysis]–[Optimization] loop. As indicated previously, the analysis (and, naturally, the sensitivity analysis) is performed with the Finite Element Method. This method, being rather time consuming, calls for a cost-effective method of sensitivity analysis. In order to fulfil also the requirement for ease of implementation, the well-known semi-analytical method as described by Cheng and Yingwei (1987) is used. It is based on a differentiation of the basic finite element equation of equilibrium:

$$[K]\{u\} = \{f\} \tag{5}$$

where $[K]$ is the global stiffness matrix, $\{u\}$ is the vector of unknown nodal displacements, and $\{f\}$ is the load vector.

Differentiation with respect to a design variable, say x_j, gives:

$$\frac{\partial [K]}{\partial x_j}\{u\} + [K]\frac{\partial \{u\}}{\partial x_j} = \frac{\partial \{f\}}{\partial x_j} \tag{6}$$

By rearranging, we find the equation for the displacement sensitivities:

$$[K]\frac{\partial \{u\}}{\partial x_j} = \frac{\partial \{f\}}{\partial x_j} - \frac{\partial [K]}{\partial x_j}\{u\} \tag{7}$$

Thus, to find the displacement sensitivities we only need to calculate $\partial\{f\}/\partial x_j$ and $\partial[K]/\partial x_j$ with a finite difference approximation and use the stiffness matrix factorization once performed in solving (6). The method is easy to implement and computationally efficient but should be used with caution in connection with elements with rotational degrees of freedom. No problems have been observed with the elements included in CAOS.

Knowing the displacement sensitivities, nodal stresses are easily found. This calculation is also based on finite differences. For each element, the vector of nodal stresses is given by

$$\{\sigma^e\} = [C][B]\{u^e\} \tag{8}$$

where $[C]$ is the constitutive matrix connecting nodal strains with nodal stresses. The matrix [C] depends only on the material characteristics and remains unchanged by a perturbation of the geometrical design variables. The matrix [B] is the geometrical condition connecting nodal displacements with strains. This matrix is a function of the element node coordinates and is expected to change with the geometry. The vector $\{u^e\}$ is the part of $\{u\}$ concerning the element in question.

Knowing the derivatives of the element nodal displacements, we can estimate nodal displacements, $\{u^e\}^*$, of the perturbed geometry:

$$\{u^e\}^* \approx \{u^e\} + \frac{\partial \{u^e\}}{\partial x_j} \Delta x_j \tag{9}$$

where Δx_j is the finite perturbation of variable x_j

Using these displacements we can calculate the stresses of the perturbed geometry directly:

$$\{\sigma^e\}^* = [C][B]^* \{u^e\}^* \tag{10}$$

where $[B]^*$ is the displacement strain matrix of the perturbed element. The stress derivative is now approximated by finite difference.

$$\frac{\partial \{\sigma^e\}}{\partial x_j} \approx \frac{\{\sigma^e\}^* - \{\sigma^e\}}{\Delta x_j} \tag{11}$$

For most problems, the calculation of stress derivatives is surprisingly stable considering the approximations involved. However, for geometries involving large rigid body rotations, the evaluation of displacement sensitivities becomes inaccurate. This, in turn, means that the stress sensitivities loose their reliability and absolute convergence of the problem may be unattainable.

6. Examples

A large number of theoretical and practical shape optimization examples have been solved with CAOS. Some of these are described by Rasmussen et. al. (1992). In this section, we present three examples of very different nature. The first example shows how CAOS can be used to improve a practical two-dimensional design significantly. The second example demonstrates how the topology/shape optimization enables the designer to very quickly arrive at a detailed solution from an initial loose idea. The third example demonstrates the three-dimensional capabilities of the CAOS system.

6.1 Engine suspension. Figure 5 shows the right half of a symmetrical engine suspension from the intercity train IC3 produced by ABB Scandia, Randers Denmark. The

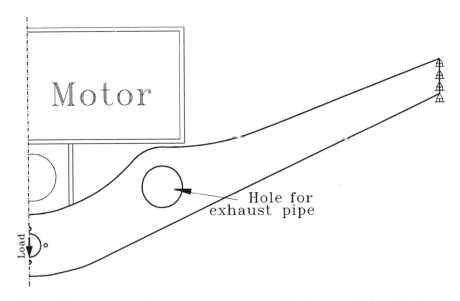

Figure 5. Right half of symmetrical engine suspension.

suspension is simply supported at its end points, i.e., vertical displacements are fixed, and the weight of the engine is transferred to the boundary of the large hole on the line of symmetry as indicated in the figure. Symmetry is imposed by fixing horizontal displacements at the symmetry boundary. The other large hole allows for passage of the exhaust pipe. Furthermore, four small holes are positioned left, right, over, and below the large hole on the symmetry line. Only the two holes above and below are considered in the finite element model.

The objective of the optimization is to minimize the volume while not exceeding the maximum allowed von Mises reference stress of 200 N/mm². We shall concentrate on modifying the external geometry and leave the sizes and shapes of the holes as they are. Furthermore, the regions around the simply supported ends have to retain their shapes due to the functionality of the suspension. For the same reason, the possible increase of thickness in the vicinity of the line of symmetry is restricted to 20 mm.

The large variation of sizes of holes calls for the use of a rather complicated design element configuration as shown in figure 6. This configuration leads to a finite element mesh that gives an acceptable modelling of the small holes while keeping the overall number of nodes at a moderate level.

The initial design has a maximum von Mises stress of 228 N/mm² and is thus infeasible by 17 %. The stress violation occurs at the boundary of the small holes at the line of symmetry. The final geometry is shown in figure 7. This geometry is feasible and the

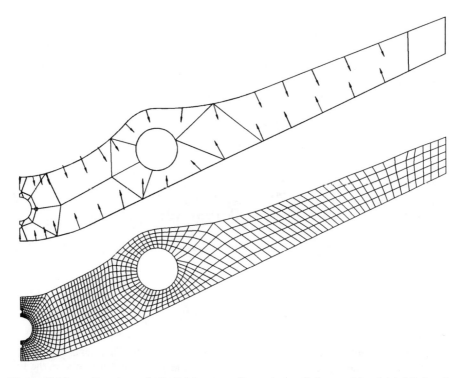

Figure 6. Move directions, design element configuration and the resulting initial finite element mesh.

volume is reduced by 29 % in comparison with the initial geometry.

6.2 A new bicycle design. As previously mentioned, CAOS has been interfaced with the HOMOPT topology optimization system by Bendsøe and Kikuchi. The topology optimization capability in connection with traditional shape optimization makes way for the use of structural optimization much earlier in the design process than with conventional shape optimization which in most cases is used only for the final details. The following example demonstrates how the combination of topology and shape optimization can be used to review a classical well-known design problem: a bicycle.

Bicycle design has evolved over many years and most professional bicycles look rather alike which leads to the suspicion that the standard bicycle frame is a highly optimized structure. Following well established bicycle design rules for a person of an approximate height of 1.90 m yields the design in figure 8 on which approximate loads are also shown. We may say that this design is the result of many years of engineering experience in bicycle design, and that future developments should be based on this foundation. Performing a topology optimization based on these loads and with a low

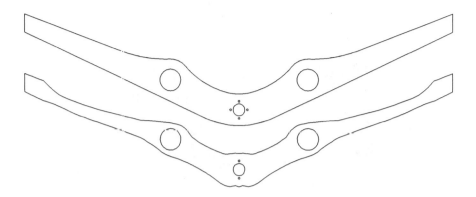

Figure 7. Comparison of initial and final engine suspension geometries. Volume reduced by 29 %.

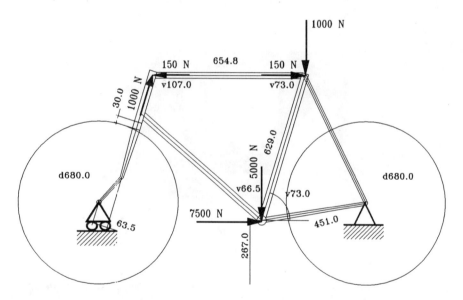

Figure 8. Classical bicycle design for a person of approximately 1.9 m height.

limit of available material, corresponding to the use of e.g. steel, yields the result of figure 9 which shows a striking resemblance with the classical frame design.

However, the advent of several new materials has led to various attempts to review the design of bicycle frames from the point of view that the use of a less stiff but also

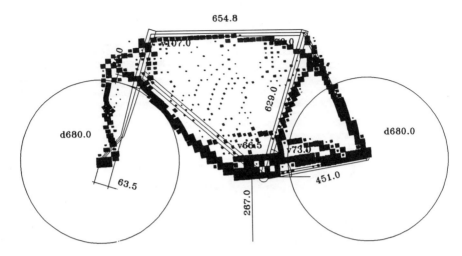

Figure 9. Result of topology optimization with a low amount of volume available.

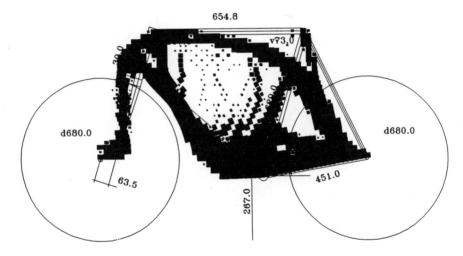

Figure 10. Result of topology optimization using twice the volume of the preceding figure.

much lighter material may lead to a better solution. Figure 10 shows the result of a topology optimization with significantly more, supposedly lighter, material available. It is obvious that this design is rather different from the classical one. Again at this stage, we bring the engineering knowledge into the process. This knowledge tells us that the shape of the front fork is important for the dynamic stability of the bicycle and therefore should retain its original design. As the whole process is performed in the interactive CAD environment, it is very easy to take such considerations into account. We therefore define a shape design model consisting only of the actual frame. We seek to minimize the volume of the bicycle without exceeding the compliance of the optimized topology of figure 10. The result of this process again can be reviewed from the engineering point of view and have practical details added, yielding the final design of figure 11.

Figure 11. Final reviewed alternative bicycle design.

6.3 Wishbone. This example deals with the optimization of a part of a front wheel suspension in a SAAB 9000 car, the so-called wishbone. This problem has previously been treated by Bråmå and Rosengren (1990) using shell elements. We shall solve the problem using 3D brick elements, and using a very complicated design model. The aim of the optimization is to reduce the weight, i.e. the volume, of the wishbone, and at the same time fulfil the two constraints mentioned below.

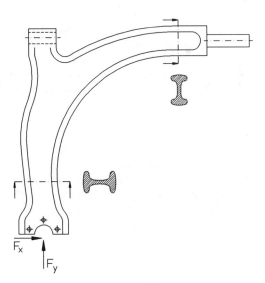

Figure 12. Initial design wishbone.

Minimize the volume
Subject to
Max. von Mises stress ≤ Initial value.
Compliance ≤ Initial value.

In Bråmå's and Rosengren's original problem, the wishbone was flexibly suspended. CAOS does not have this facility, and the result of the two optimizations can therefore not be directly compared. This is also the reason why we choose initial values as constraint limits. In order to avoid undesirable stress concentrations in the actual geometry, small pieces of extra material have been added to the model at the points of mechanical fixation.

Figure 12 shows the initial design of the wishbone. In the definition of the problem it will be assumed that the upper left and right boundaries are fixed as indicated in figure 13. The loads acting on the wishbone will be defined in three different load cases corresponding to the following situations:

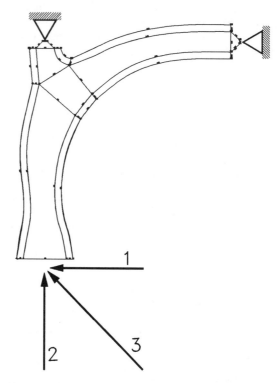

Figure 13. Load cases and boundary conditions.

1. Maximum straight line braking, i.e., F_x is non zero.
2. Maximum lateral acceleration, i.e., F_y is non zero.
3. Maximum combined braking/lateral acceleration, i.e., both F_x and F_y are non zero.

Figure 14 illustrates the rather complex CAD model. The geometry is modelled by 202 boundaries, 30 design element and 308 move directions. The number of independent design variables is reduced to 112 by linking some of the move directions. It is evident that the user has to rely heavily on good interactive graphical facilities to maintain the overview of a model as complicated as this. The lengths of the move directions in figure 14 also indicate the geometrical constraint that, due to physical limitations, the wishbone has to remain within a predetermined space.

The wishbone is made of aluminum and has the following material parameters:

Young's modulus: 71000 MPa.
Poisson's ratio: 0.3.

The CAOS system

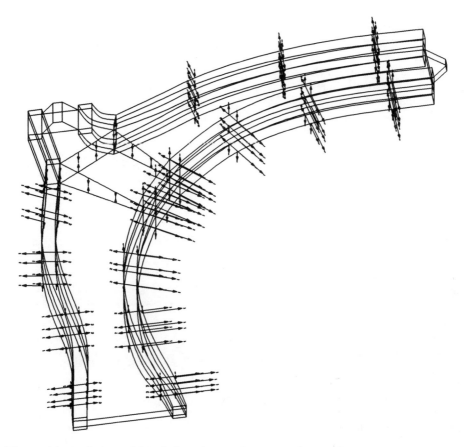

Figure 14. Analysis model including design elements and move directions.

The initial finite element mesh is illustrated in figure 15 and is modelled by 776 8-node isoparametric solid elements.

The optimization results in a considerable volume reduction of 40% over a total of 9 iterations. The final finite element model, which is feasible for all load cases, is shown in figure 16. In spite of the considerable volume reduction, the visual geometrical alterations are rather small.

7. Future work

When the development of CAOS was initiated in 1986 the introduction of the facilities of the present system was not anticipated. The simple numerical part of the system was programmed in FORTRAN77 and the more complicated parts involving mesh generation, parameterization of the geometry, manipulation of geometrical entities etc. were coded in Pascal, which is better suited for the storage and manipulation of complicated data structures.

The expansion of CAOS to three dimensions and the desire to steadily improve the analysis facilities has created the need for a revision of the basic data structures of the system to better cope with a multitude of different problem formulations and analysis methods. Furthermore, the bilingual code made the system less portable.

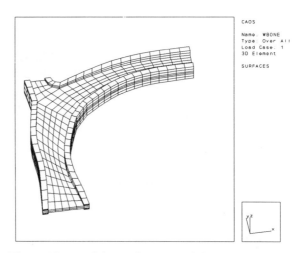

Figure 15. Initial finite element model.

This led to the decision of creating a new foundation for a system based on the same philosophy as CAOS but with a more systematic organization and much larger potential for extension into new problem fields such as dynamic problems, shell structures, non-linear analysis, anisotropic problems, etc. The planning and coding of this new system, called the Optimum Design System (ODESSY) was initiated in

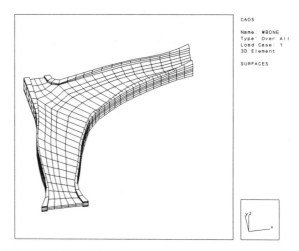

Figure 16. Final finite element mesh.

the Summer of 1991. It will be coded entirely in ANSI C and C++. However, the CAOS system will contiuously be developed and improved as new research results become available.

References

N. Olhoff and J.E. Taylor (1983), On structural optimization. Journal of Applied Mechanics, Vol. 50, pp. 1139-1151.

Björn J.D. Esping (1984), The OASIS Structural Optimization System. Computers & Structures, Vol. 23, pp. 365–377.

V. Braibant and C. Fleury (1984), Shape Optimal Design using B–splines. Computer Methods in Applied Mechanics and Engineering, vol. 44, pp. 247 – 267.

J.A. Bennett and M.E. Botkin (1985), Structural Shape Optimization with Geometric Description and Adaptive Mesh Refinement. AIAA Journal, vol. 23, no. 3, pp. 458 – 464.

M.E. Botkin, R.J. Yang and J.A. Bennet (1986), Shape Optimization of Three–dimensional Stamped and Solid Automotive Components. James A. Bennet and Mark E. Botkin (ed): The Optimum Shape. Automated Structural Design, Plenum Press, New York.

E.L. Stanton (1986), Geometric Modeling for Structural and Material Shape Optimization. James A. Bennet and Mark E. Botkin (ed): The Optimum Shape. Automated Structural Design, Plenum Press, New York.

Yunliang Ding (1986), Compendium. Shape Optimization of Structures: A Literature Survey. Computers and Structures, vol. 24, no. 6, pp. 985–1004.

H.R.E.M. Hörnlein (1987), Take–Off in Optimum Structural Design. C.A. Mota Soares (ed.): Computer Aided Optimal Design. Structural and Mechanical Systems. NATO ASI Series F: Computer and System Sciences, Vol. 27, Springer Verlag.

Martin P. Bendsøe (1989), Optimal shape design as a material distribution problem. Structural Optimization, vol. 1, no. 4, pp. 193–202.

Niels Olhoff (1989), Multicriterion structural optimization via bound formulation and mathematical programming. Structural Optimization, vol. 1, no. 1, pp. 11–17.

K. Svanberg (1987), The Method of Moving Asymptotes – a new method for structural optimization. International Journal for Numerical Methods in Engineering, vol. 24, pp. 359–373.

G. Cheng and L. Yingwei (1987), A new computation scheme for sensitivity analysis. Engineering Optimization, Vol. 12, pp. 219–234.

John Rasmussen, Erik Lund and Torben Birker (1992), Collection of Examples, CAOS Optimization System, 3rd edition. Special Report no. 1c, Institute of Mechanical Engineering, Aalborg University. Available from the authors at the cost of printing.

Helder C. Rodrigues (1988), Shape optimal design of elastic bodies using a mixed variational formulation. Computer Methods in Applied Mechanics and Engineering, vol. 69, pp. 29–44.

Torsten Bråmå and Ragner Rosengren (1990), Applications of the structural optimization program OPTSYS, Proceeding of the ICAS Conference, ICAS-90-2.1.3.

Authors' addresses:

John Rasmussen, Erik Lund and Niels Olhoff
Institute of Mechanical Engineering
Aalborg University
Pontoppidanstraede 101
DK-9220 Aalborg East
Denmark.
Tel +45 98 15 85 22
Fax +45 98 15 14 11
E-mail i15jr@aud.auc.dk

Torben Birker
Universität-GH Essen
Fachbereich 10, Bauwesen,
Postfach 10 37 64
D-4300 Essen 1,
Germany.
Tel +49 0201 183 2695
Fax +49 0201 183 2201

Shape Optimization with Program CARAT

Kai–Uwe Bletzinger, Reiner Reitinger, Stefan Kimmich und Ekkehard Ramm

Abstract. Shape optimal design is the synthesis of structural analysis and mathematical programming combined with computer aided geometric design (CAGD) concepts and behavior sensitivity analysis. This generally accepted idea is realized by CARAT. The underlying models are formulated with special regard to a general overall model of structural optimization which is efficient as well as flexible enough to be applied to shape optimal design of arbitrary shells in three dimensional space. Special emphasis is given to potential optimization algorithms (e.g. an extension of the method of moving asymptotes), practical design capabilities, clearly formulated interactions like variable linking, and modular computer codes. Shape optimizations of an initially axisymmetric shell, a fly–wheel and a bell demonstrate the capabilities of the presented approach.

1. Introduction

The methods of structural optimization made great progress within the last decade and are now well on the way to be accepted as advanced engineering tools for the design of structures. A lot of special software and procedures have been developed. Some of them are presented in this monograph. The related process is reviewed and accompanied by a number of excellent papers (Schmit 1981, Vanderplaats 1982, Haftka and Grandhi 1986), proceedings (Bennett and Botkin 1986, Mota Soares 1987, Eschenauer und Thierauf 1989a, Rozvany 1991), and books (Vanderplaats 1984, Atrek et al. 1984, Haftka et al. 1990) which clearly show the evolution from the first beginnings with sizing of structures to the latest developments in shape optimal design. Only some can be cited. At present, structural optimization is understood as a synthesis of several individual disciplines which individually are to a large extent developed: (i) design modeling, (ii) structural and behavior sensitivity analysis, and (iii) mathematical programming. This idea (presented e.g. by Braibant and Fleury 1986) appeared to be the basic concept for the development of effective software. It is adopted by all the leading researchers and software developers and became known e.g. as "three column concept" (Eschenauer et al. 1989b). CARAT is organized the same way (Bletzinger et al. 1992a). The main motivation was to develop a broad and flexible program basis which allows to develop and apply new algorithms for research purposes very easily. This is reflected by the name: *Computer Aided Research Analysis Tool*.

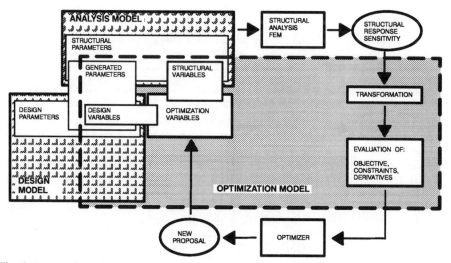

Fig. 1. *Interaction of optimization, design and analysis models.*

The backbone of CARAT is a powerful data base which allows to store and to address arrays and parameters by names and paths in a complete tree–like data structure, although the whole package is written in FORTRAN 77 because of traditional and compatibility reasons. This strict data organization allows several distinct lines of research and to maintain previous developments at the same time: non–linear structural behavior, limit state and stability analysis, dynamic behavior, finite element formulations, material modeling, adaptivity, pre– and post–processing, behavior sensitivity analysis, and nonlinear optimization algorithms. The special aim of applications in the field of shape optimization are free formed shell structures in civil and mechanical engineering including axisymmetric structures, plates and membranes as special cases (Ramm et al. 1991a, Ramm and Mehlhorn 1991b, Bletzinger and Ramm 1992b). The structures are discretized by isoparametric degenerated displacement elements; linear response is assumed. Nevertheless, it has been shown that the ultimate load and behavior of concrete shells can be tremendously improved by shape optimization with respect to linear behavior when after this procedure the optimized structures were subjected to a geometrically and materially non–linear analysis (Ramm and Reitinger 1992). The underlying typical design loop is sketched in Fig. 1 and reflects the interactions of the different modelling stages.

The contribution will give some insight into the formulation and realization of structural optimization problems with CARAT. Special focus is laid on the formulation of the involved model stages and their interactions. Some examples will finish the contribution.

2. Optimization Model

The general non-linear optimization problem can be stated as:

$$\begin{aligned}
\text{minimize} \quad & f(x) \\
\text{subject to:} \quad & g_j(x) = 0 \; ; \; j = 1, ..., m_e \\
& g_j(x) \leq 0 \; ; \; j = m_e + 1, ..., m \\
& \underline{x} \leq x \leq \bar{x} \; ; \; x \in \mathbb{R}^n
\end{aligned} \quad (1)$$

Objective and constraints for one and more load cases as they are available in CARAT are given in Table 1. General objectives can be defined by the weighted sum of single objectives to consider multi–criterion problems by the relation:

$$f(x) = \sum_{i=1}^{k} w_i f_i(x) \qquad \begin{array}{l} w_i \; ... \; \text{weighting factor} \\ k \; ... \; \text{number of distinct objective functions} \end{array} \quad (2)$$

The optimization problem can be tackled either directly or by successive solutions of approximated subproblems. Different schemes render problem oriented approximations:

Design variables x:

r_d, r_a	Design and structural nodes
t_d, t_a	Design and structural thicknesses
A	other cross section parameters

Objective functions f(x):

$f_W = \int \varrho \, dv$ — Weight or volume

$f_E = \frac{1}{2} \int \sigma \varepsilon \, dv$ — Strain energy

$f_S = \int (\sigma - \sigma_{avg})^2 \, da$ — Stress leveling

$f_C = \int \varrho r^2 \, dv$ — Rotational inertia

$f_F = a_i$ — Frequency

$f_T = \sum_i \dfrac{(a_i - a_{i0})^2}{a_{i0}^2} w_i$ — Frequency tuning with weighting factor w_i

Constraint functions g(x):

$g_W = \dfrac{W}{W_{all.}} - 1 = 0$ Weight or volume

$g_C = \dfrac{C}{C_{all.}} - 1 = 0$ Rotational inertia

$g_u = \dfrac{u}{u_{all.}} - 1 \leq 0$ Displacement

$g_\sigma = \dfrac{\sigma}{\sigma_{all.}} - 1 \leq 0$ Stresses

$g_F = \dfrac{a_i}{a_{i0}} - 1 = 0$ Frequency

Table. 1. *Typical variables and functions in optimal shape design of shells.*

- Hybrid function approximation (Starnes and Haftka 1979): first order approximation of problem functions with respect to either the original variables x_i or their inverse $1/x_i$ depending on the sign of partial derivatives.
- Method of moving asymptotes (MMA, Svanberg 1987): approximation of problem functions with respect to variables

$$y_i = \frac{1}{(x_i - L_i)} \quad \text{or} \quad y_i = \frac{1}{(U_i - x_i)} \tag{3}$$

depending on the sign of partial derivatives. L_i and U_i are lower and upper asymptotes, respectively. The original method is extended to consider unbounded problems and equality constraints. More details in the next chapter.

- General power approximation (GPA): extension of MMA considering the ideas of Prasad (1983). Functions are approximated with respect to variables

$$y_i = -\frac{1}{b}(x_i - L_i)^b \quad \text{or} \quad y_i = -\frac{1}{b}(U_i - x_i)^b \tag{4}$$

respectively. Exponents can be chosen explicitly or determined by second order derivatives which are estimated by forward differences.

All approximation schemes can be applied selectively for any function and any variable by extended input statements which allow efficient problem generation. Satisfactory results have been achieved for all kinds of general problems by MMA with bounded approximation of the objective function.

Five different optimizers are available:

- NLPQL (Schittkowski 1981): sequential quadratic programming method (SQP) based on direct, iterative solution of the nonlinear Kuhn–Tucker conditions. The Hessian of the Lagrangian function is approximated by BFGS recursive formula. The method is generally applicable for unconstrained as well as equally and non–equally constrained problems. Very good experiences are made if NLPQL was applied directly to all kinds of complex problems of shape optimal design.
- SLP: an own development of the well known sequential linear programming algorithm. Simplex algorithms from standard program libraries can be used to solve the linear sub–problems. The method behaves well for sizing problems but cannot be recommended for shape optimization since unbounded subproblems may be generated. Convergence is very much dependent on move limit adaptation. The standard L_1–penalty function is introduced as merit function which is used in a line search procedure to adjust step sizes.

- DUAL: conjugate gradient algorithm extended to consider simple side constraints on variables to solve separably approximated sub–problems. The algorithm is used only if minimization of sub–problems for primal variables can be done explicitly (Fleury 1986). Otherwise NLPQL is used to ensure stable and efficient solutions.
- EVOL (Schwefel 1981): two member evolution strategy. A direct search method with randomly generated step sizes and directions. Interesting method for discontinuous problems which has further yet unexploited potential for parallel processing or topology optimization.
- FSD (Gallagher 1973): fully stressed design algorithm used for sizing of trusses and plates by the stress ratio method. Elasto–plastic buckling of single members due to German steel design code DIN 18800 is considered by a spline approximation. Also used for optimal material distribution to generate structural topologies.

All optimizers can be used successively in one optimization run by indicating a so called "optimization path". Special potentials can be exploited, e.g. by a few steps of FSD or MMA followed by more precise but costly optimization with NLPQL (Fig. 2).

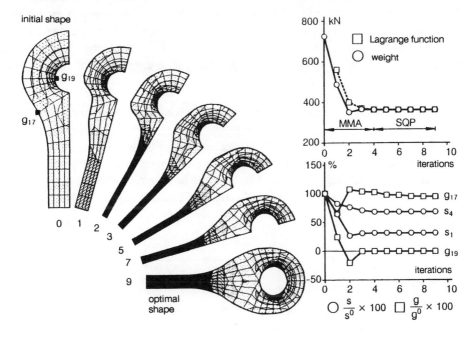

Fig. 2. *Connecting rod: iteration history (MMA and NLPQL).*

3. An Extension of the Method of Moving Asymptotes

The method of moving asymptotes (Svanberg 1987) is derived from a problem linearization with respect to generalized variables y:

$$\tilde{t}(y)^{(k)} = t(y^{(k)}) + \left(\frac{\partial t(y^{(k)})}{\partial y}\right)^T \left(y^{(k+1)} - y^{(k)}\right) \tag{5}$$

$$y_i = \begin{cases} \dfrac{1}{U_i^{(k)} - x_i} & ; \quad \dfrac{\partial t_j(x^{(k)})}{\partial x_i} \geq 0 \\[2ex] \dfrac{1}{x_i - L_i^{(k)}} & ; \quad \dfrac{\partial t_j(x^{(k)})}{\partial x_i} < 0 \end{cases} \tag{6}$$

where t is any problem function (objective or constraint), $\tilde{t}^{(k)}$ its explicit approximation in the k-th iteration step, $x^{(k)}$ the point of expansion, and $L_i^{(k)}$ and $U_i^{(k)}$ are lower and upper asymptotes of variable x_i, respectively. The asymptotes are adopted by a heuristic rule to adjust the curvature of the approximation which gave the method its name. The choice of (5) and (6) generates convex and separable sub–problems which can be solved efficiently by dual methods.

Nevertheless, MMA suffers from two drawbacks: (i) it is based on linearization which generates unbounded sub–problems in the case of little or even unconstrained problems as they often appear in shape optimal design, and (ii) the heuristic adaptation of asymptotes may cause convergence problems near the optimum or may eventually even lead to divergence. By a simple extension of Svanberg's idea both drawbacks can be overcome while all the well known advantages of the method are preserved (Bletzinger 1992c). Now, the objective function $f(x)$ is approximated with respect to both asymptotes simultaneously:

$$\tilde{f}(x)^{(k)} = r_0^{(k)} + \sum_{i=1}^{n} \left(\frac{p_{0i}^{(k)}}{U_i^{(k)} - x_i} + \frac{q_{0i}^{(k)}}{x_i - L_i^{(k)}} \right) \tag{7}$$

$$p_{0i}^{(k)} > 0 \quad \text{and} \quad q_{0i}^{(k)} > 0$$

Inequality constraints are approximated as suggested by Svanberg:

$$\tilde{g}_j(x)^{(k)} = r_j^{(k)} + \sum_{i=1}^{n} \left(\frac{p_{ji}^{(k)}}{U_i^{(k)} - x_i} + \frac{q_{ji}^{(k)}}{x_i - L_i^{(k)}} \right) \quad ; \quad j = m_e + 1 \ldots m \tag{8}$$

where $p_{ji}^{(k)} > 0$ and $q_{ji}^{(k)} = 0$

or $p_{ji}^{(k)} = 0$ and $q_{ji}^{(k)} > 0$ depending on the sign of partial derivatives.

Additionally, the method is extended for equality constraints. By use of the following approximation the sub–problem is still convex at the Kuhn–Tucker point and can be solved explicitly:

$$\tilde{g}_j(x)^{(k)} = r_j^{(k)} + \sum_{i=1}^{n} \left(\frac{p_{ji}^{(k)}}{U_i^{(k)} - x_i} - \frac{p_{ji}^{(k)}}{x_i - L_i^{(k)}} \right) \quad ; \quad j = 1 \ldots m_e \quad (9)$$

The asymptotes are determined by an approximation of the first Kuhn–Tucker condition, comparable to the strategy of Lagrange methods: the gradient of the Lagrangian function Φ and an estimation of its curvature at $x^{(k)}$ are used. The asymptotes $L_i^{(k)}$ and $U_i^{(k)}$ are assumed to be generated by the rule:

$$L_i^{(k)} = x_i^{(k)} - \Delta_i^{(k)} \quad \text{and} \quad U_i^{(k)} = x_i^{(k)} + \Delta_i^{(k)} \quad (10)$$

The increments $\Delta_i^{(k)}$ are evaluated as:

$$\Delta_i^{(k)} = \frac{1 + (ca_i)^2}{ca_i} \frac{\Phi'_i}{\Phi''_i} \quad (11)$$

The factor a_i is defined with respect to the sign of the Lagrangian's gradient:

$$a_i = \begin{cases} -\frac{1}{\Phi'_i}\left(A_i - \sqrt{A_i^2 - \Phi'^2_i}\right) & ; \quad \text{if } A_i \le \Phi'_i < 0 \\ \frac{1}{\Phi'_i}\left(B_i - \sqrt{B_i^2 - \Phi'^2_i}\right) & ; \quad \text{if } B_i \ge \Phi'_i > 0 \\ 1 & ; \quad \text{else} \end{cases} \quad (12)$$

where A_i and B_i are:

$$A_i = \sum_{j=1}^{m_e} v_j g'_{ji} + 2 \sum_{j=m_e+1}^{m} \max(0, v_j g'_{ji}) - \Phi'_i \quad (13)$$

$$B_i = \Phi'_i - \sum_{j=1}^{m_e} v_j g'_{ji} - 2 \sum_{j=m_e+1}^{m} \min(0, v_j g'_{ji}) \quad (14)$$

( )$'_i$ denotes partial derivative with respect to x_i at $x^{(k)}$. The curvature of the Lagrangian is estimated by a finite difference step:

$$\Phi''_i = \Phi''^{(k-1)}_i = \begin{cases} \max \left\{ \begin{array}{l} \dfrac{\Phi'_i(x^{(k)}, v^{(k)}) - \Phi'_i(x^{(k-1)}, v^{(k)})}{x_i^{(k)} - x_i^{(k-1)}} \\ \alpha\, \Phi''^{(k-1)}_i \end{array} \right\} \quad ;\ k > 0 \\ 1 \qquad\qquad\qquad\qquad\qquad\qquad\qquad\qquad ;\ k = 0 \end{cases} \tag{15}$$

The Lagrange multipliers are denoted by v_j. Two additional parameters are introduced: (i) a relaxation factor c which allows to vary the approximation between the original linearization according to Svanberg ($c = 1$) and diagonal quadratic expansion ($c \to 0$), (ii) a factor α which preserves the curvature from becoming negative or dropping too rapidly. Satisfactory results have been achieved with $\alpha = 0.2$ which is according to the experiences with quasi Newton type Lagrange methods. With the above expressions the remaining coefficients are determined as (where all derivatives are evaluated at $x^{(k)}$):

$$p_{ji}^{(k)} = \tfrac{1}{2} \Delta_i^{2(k)} \frac{\partial g_j}{\partial x_i} \qquad ;\ j = 1\ \ldots\ m_e \tag{16}$$

$$p_{ji}^{(k)} = \begin{cases} \Delta_i^{2(k)} \dfrac{\partial g_j}{\partial x_i} & ;\ \text{if } \partial g_j/\partial x_i > 0 \\ 0 & ;\ \text{else} \end{cases} \quad ;\ j = m_e + 1\ \ldots\ m \tag{17}$$

$$q_{ji}^{(k)} = \begin{cases} 0 & ;\ \text{if } \partial g_j/\partial x_i \geq 0 \\ -\Delta_i^{2(k)} \dfrac{\partial g_j}{\partial x_i} & ;\ \text{else} \end{cases} \quad ;\ j = m_e + 1\ \ldots\ m \tag{18}$$

$$p_{0i}^{(k)} = \tfrac{1}{2} \Delta_i^{2(k)} \left(\tfrac{1}{2} \Delta_i^{(k)} \Phi''_i + \Phi'_i \right) - \sum_{j=1}^m v_i^{(k)} p_{ji}^{(k)} \tag{19}$$

$$q_{0i}^{(k)} = \tfrac{1}{2} \Delta_i^{2(k)} \left(\tfrac{1}{2} \Delta_i^{(k)} \Phi''_i - \Phi'_i \right) - \sum_{j=1}^m v_i^{(k)} p_{ji}^{(k)} \tag{20}$$

The following academic but typical problem shows the potential of the proposed extension. Shape and cross sections of the three bar truss (Fig. 3) are to be determined. The objective to

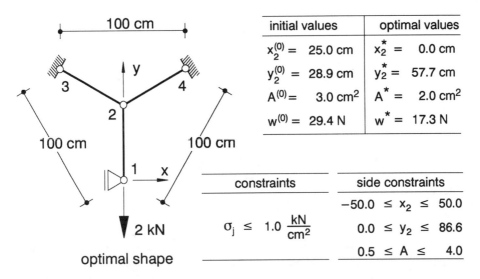

Fig. 3. *Another 3–bar truss problem.*

be minimized is structural weight. Variables are the location x_2, y_2 of the center node and the common cross section A of all bars. The truss is loaded by a vertical force where the tensile stresses are limited to 1.0 kN/cm^2. At the optimum the variables are uncoupled. This means that the original method converges only if the asymptotes are adapted, eventually leading to failure. The adaptation rule by Svanberg with an adaptation factor of $s = 0.3$ was used. The extended method appears to be substantially more stable as can be seen in Fig. 4.

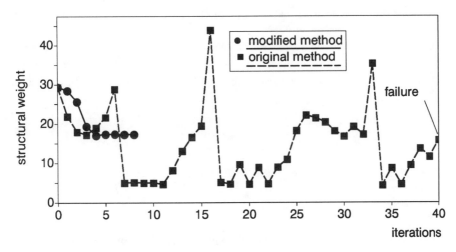

Fig. 4. *Three bar truss: iteration histories.*

4. Design Model

The general methods of Computer Aided Geometrical Design (CAGD, Faux and Pratt 1979, Böhm et al. 1984) are the basis of the CARAT design model. Shapes are approximated piecewise by "design elements". Within each design element the resulting shape r_a is parametrized in terms of shape functions H_i, element parameters (u, v, w), and design nodes r_{di} which describe the location of the element in space:

$$r_a(u, v, w) = \sum_{i=1}^{n} H_i(u, v, w)\, r_{di} \qquad (21)$$

There are many different shape functions available (Fig. 5). Geometries and tangential support conditions can be generated in Cartesian, cylindrical, and spherical coordinates as well as their analytical derivatives with respect to the optimization variables. This is very important in shape optimal design but quite a unique feature of CARAT design (Bletzinger 1990). For the shape design of free formed shells cubic Bézier– and B–splines, and two dimensional bi–cubic Bézier patches appear to be superior to others.

Shapes of realistic structures are usually modelled by more than one design element. Especially in the case of free formed surfaces it is important to consider continuity requirements at design element interfaces. To get sufficiently smooth shapes it is necessary to force at least continuous slopes whereas continuous curvatures improve the impression of smoothness even more. Since continuity requirements apply for the generated (i.e. visual) shape and have to be independent of curve or surface parametrizations they are called "visual" or "geometrical" continuity requirements (Böhm et al. 1984). Because of their special construction C_1–geometrical or G_1–continuity between Bézier– and B–splines can be generated in a very elegant manner. It is sufficient to force the common node and the next control nodes on either side to stay on a common line (Fig. 6). This additional constraint in terms of structural optimization is expressed by the relation:

$$3\,\delta\left(r_n^{(1)} - r_{n-1}^{(1)}\right) = 3\left(r_1^{(2)} - r_0^{(2)}\right) \qquad (22)$$

$$\text{with} \qquad r_n^{(1)} = r_0^{(2)}$$

where $r_i^{(1)}$ and $r_i^{(2)}$ are the position vectors of control nodes of the two adjacent splines. δ is a free parameter which will be called "tangent factor".

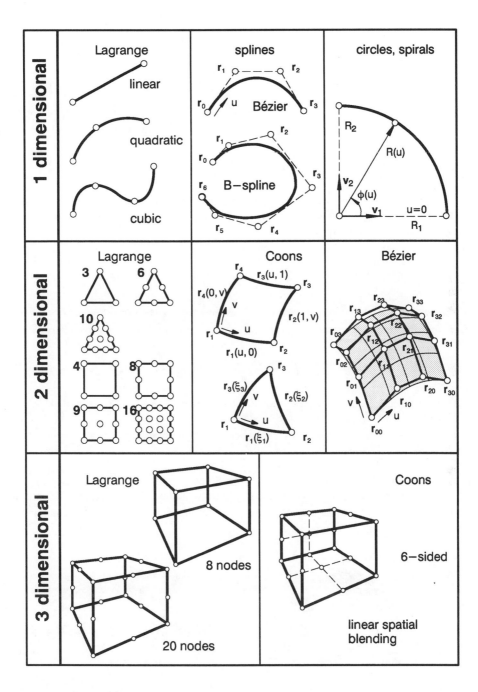

Fig. 5. *Design elements available in CARAT design concept.*

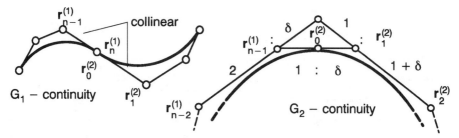

Fig. 6. *Geometrical continuity of Bézier– and B–splines.*

It is convenient to fulfill (22) explicitly. This will reduce the number of optimization variables and will speed up and stabilize the optimization procedure. $r_1^{(2)}$ can be expressed in terms of the two other nodes and δ (Fig. 6):

$$r_1^{(2)} = r_0^{(2)} + \delta \, (r_n^{(1)} - r_{n-1}^{(1)}) \qquad (23)$$

An equivalent rule which ensures continuous curvature between two B–splines affects in total five control nodes: the common node and the two next nodes of each connected curve.

The relations state as (compare with Fig. 6):

$$r_{n-1}^{(1)} = \frac{1}{(1+3\delta)(2+\delta)} \left[\delta \, (1+\delta) \, r_{n-2}^{(1)} + 2 \, (1+\delta)(2+\delta) \, r_n^{(1)} - 2 \, r_2^{(2)} \right]$$

$$r_1^{(2)} = \frac{1}{(1+3\delta)(2+\delta)} \left[-\delta^2(1+\delta) \, r_{n-2}^{(1)} + (1+\delta)^2 \, (2+\delta) \, r_0^{(2)} + 2 \, \delta \, r_2^{(2)} \right] \qquad (24)$$

G_1–continuity of Bézier patches is enforced by linear relations between at most nine control nodes. The location of five dependent nodes can be described in terms of four remaining free nodes and two so called "line factors" δ_1 and δ_2.

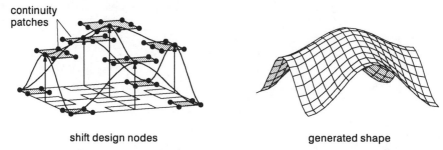

Fig. 7. *Interactive surface modification; continuity patches connecting four Bézier patches.*

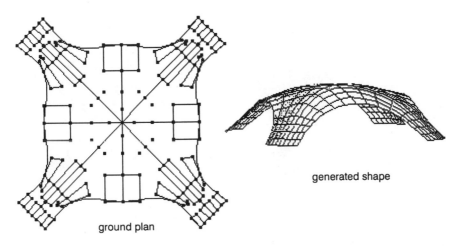

Fig. 8. *Free formed shell (16 Bézier elements).*

All these topological dependencies are preserved by superimposed "tangent elements" (1–D case) and "continuity patches" (2–D case). On request they are generated automatically by CARAT and are preserved during interactive user manipulations and shape optimization. Fig. 7 shows different types of continuity patches depending on whether they are connecting two or four design patches or are defined at an isolated corner. In either case four nodes are independent and control the locations of the remaining nodes leading to a reduction of geometrical degrees of freedom which is very welcome in structural optimization to stabilize the procedure.

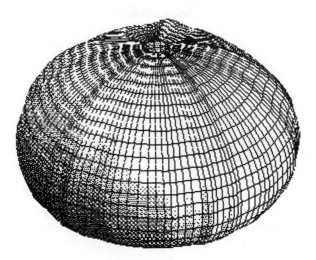

Fig. 9. *Finite Element model of a sea urchin shell.*

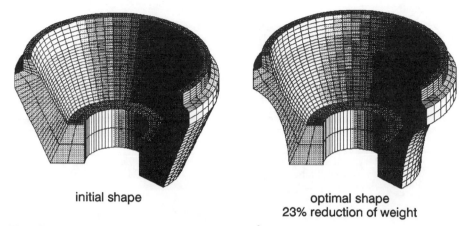

initial shape

optimal shape
23% reduction of weight

Fig. 10. *Shape optimization of a connection cone[1].*

The idea of continuity patches is very helpful in interactive design of free formed shells which can serve as initial shapes for subsequent optimization runs or as valuable interactive pre–processor tool for input preparation of complex shapes. Figure 8 shows the plan of a free formed shell described by in total 16 Bézier patches and the generated shape modeled by 8–noded isoparametric shell elements. The generated model of a sea urchin shell is shown in Fig. 9 which was the subject of a biomechanical study together with biologists (Philippi and Nachtigall 1991). The stiffening effects of the wrinkles were the main objective of this investigation. The shape was interactively adjusted to measured data using B–splines which were linearly blended in cylindrical coordinates (Coons interpolation).

Fig. 10 shows the use of CARAT design for 3D visualization of an axisymmetrical shape optimization problem.

5. Sensitivity Analysis

The sensitivity analysis supplies gradient informations on objective $f(x)$ and constraints $g(x)$ with respect to optimization variables. In general, any function t (objective or constraint) depends on optimization variables x and state variables u, e.g. displacements. Thus, the total derivative of t with respect to x is given as:

$$\frac{dt}{dx} = \frac{\partial t}{\partial u}\frac{du}{dx} + \frac{\partial t}{\partial x} \tag{25}$$

where the determination of the behavior sensitivity du/dx is part of the job. It can be done by several different techniques. They can be divided into variational or discrete methods, depen-

---

[1] together with delta–X Ingenieurgesellschaft GmbH, Stuttgart.

dent on whether the gradients are obtained before or after discretization. Nevertheless, the same results will be obtained if variation, discretization, and derivation are done consistently (Kimmich 1990). In CARAT all three variants of discrete sensitivity analysis (DSA) are available: numerical, semi–analytical, and analytical. For further details refer to the extensive literature on the subject (e.g. Haftka et al. 1990, Haftka and Adelman 1989). Only some remarks on the analytical derivation of finite element matrices will be given and some own experiences reported. More informations on variational approaches may be found in Haug et al. (1986).

As it is well known the analytical derivation of the state variables is given as:

$$\frac{du}{dx} = K^{-1}\left(\frac{dR}{dx} - \frac{dK}{dx}u\right) \quad (26)$$

where R is the load vector and K is the stiffness matrix. The major concern is the calculation of the pseudo load vector

$$\overline{R} = \frac{dR}{dx} - \frac{dK}{dx}u \quad (27)$$

which results in analytical derivations of R and K. In the special case of isoparametric finite elements the derivation of an element stiffness matrix yields:

$$k_e = \int B^T C \ B \ |J| \ d\xi \ d\eta \quad (28)$$

$$k_{e,x} = \int B_{,x}^T C \ B \ |J| \ d\xi \ d\eta + \int B^T \ C_{,x} \ B \ |J| \ d\xi \ d\eta$$

$$+ \int B^T C \ B_{,x} \ |J| \ d\xi \ d\eta + \int B^T \ C \ B \ |J_{,x}| \ d\xi \ d\eta \quad (29)$$

where B is the kinematic operator, C the constitutive matrix, and $|J|$ the determinant of the Jacobian matrix. The derivative with respect to the optimization variables (size or shape) is denoted by $(\ )_{,x}$. ξ and η are the natural coordinates of the finite elements.

The analytical version of DSA appears to be very reliable for all kinds of applications at the cost of a higher programming effort. For problems with a large number of state variables u it is far more efficient than the numerical forward difference scheme because the stiffness matrix has to be factorized only once for the derivation with respect to all variables. An additional speed up occurs, if only parts of the structure are affected by shape variations. Otherwise, since the calculation of the stiffness derivative needs about four times longer than the evaluation of K itself, numerical DSA may be the right choice for problems with comparatively small number of state variables u. Within the integrated in–core structure of CARAT very small difference steps can be chosen and, therefore, no appreciable numerical differences between the approaches are observed. It definitely depends on the optimization problem which method is to be preferred; therefore all of them should be provided (Kimmich 1990).

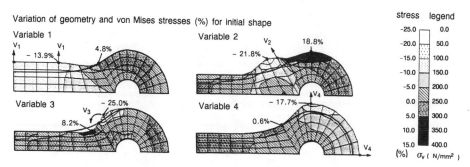

Fig. 11. *Connecting rod: v. Mises effective stress sensitivity coefficients.*

Besides the evaluation of derivatives in the context of structural optimization sensitivity data are valuable informations to rate the structural behavior and quality. The designer can use this information to identify relevant structural parameters for his own decisions or the formulation of a well posed optimization problem. CARAT supplies with capabilities to display geometric variation and stress sensitivity coefficients simultaneously (Fig. 11).

6. Model Linking

The complexity of shape optimal design becomes obvious if we take a look at the variable interactions of the different models. A commonly used rule which links variables r_a of the analysis model via the design model variables r_d with variables x of the optimization model is defined as (Bletzinger 1990):

$$r_a = r_a^0 + L_{ax}x + H_{ad}(r_d) \tag{30}$$

with:
$$r_d = r_d^0 + L_{dx}x \tag{31}$$

In these relations r_a^0 and r_d^0 denote coordinates of analysis and design models, respectively, which remain constant during the optimization process. Linking matrices L_{ax} and L_{dx} describe linear relations between optimization variables x and variable coordinates of analysis and design model, respectively. H_{ad} denote nonlinear relations between design and analysis model and are identified as shape functions. In the simple case of Cartesian coordinates shape functions define linear relations. Then H_{ad} reduces to:

$$H_{ad}(r_d) = H(u,v,w)\, r_d \tag{32}$$

where u, v, w are parameters describing the shape (compare with equation 21).

If cylindrical or spherical coordinates are used to describe the design model an additional transformation is necessary to generate final Cartesian coordinates. Now H_{ad} are nonlinear functions of r_d. They can be decomposed as:

$$H_{ad}(r_d) = T \left[H(u, v, w) \, r_d \right] \tag{33}$$

where $T[z]$ denote nonlinear transformations.

In the case of cylindrical coordinates $z^T = (\varrho, \theta, \zeta)^T$, $T[z]$ is defined as:

$$T[z] = \begin{Bmatrix} \varrho \cos \theta \\ \varrho \sin \theta \\ \zeta \end{Bmatrix} \tag{34}$$

The coefficients of linking matrix L_{dx} are defined by linking rules which help to fix typical and problem oriented degrees of freedom. Among them the following rules are useful in practical applications (Fig. 12):

- prescribed move direction: $\quad r_d = r_d^0 + s \, x_i$ \hfill (35)

- linear combination: $\quad r_d = r_d^0 + \sum_i \alpha_i \, s_i \, x_i$ \hfill (36)

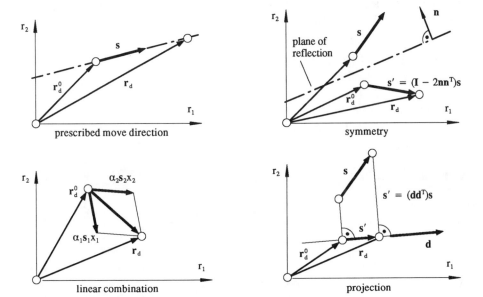

Fig. 12. *Some linear design node linking rules.*

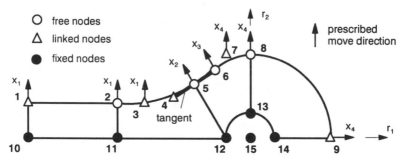

Fig. 13. *Connecting rod: linear linking rules.*

- change of basis: $$r_d = r_d^0 + s_1\, x_1 + s_2\, x_2 + s_3\, x_3 \quad (37)$$

- symmetry: $$r_d = r_d^0 + (I - 2\, nn^T)\, s\, x_i \quad (38)$$

- projection: $$r_d = r_d^0 + (dd^T)\, s\, x_i \quad (39)$$

where s_i are the vectors of movement or base directions, respectively. n is the normal vector on the plane of symmetry, and d is the direction on which s is projected. Continuity conditions as they have been introduced before are another example of linear linking.

With equations (30) and (32) behavior sensitivity of any structural response g which is originally defined in terms of analysis variables r_a and state variables u is now defined as:

$$\frac{dg(u, r_a)}{dx} = \left(\frac{\partial g}{\partial r_a} + \frac{\partial g}{\partial u}\frac{du}{dr_a} \right)\frac{dr_a}{dx} \quad \text{with} \quad \frac{dr_a}{dx} = L_{ax} + H\, L_{dx} \quad (40)$$

To give an example some linking rules are introduced for the connecting rod problem to reduce the number of variables to four (Fig. 13).

7. Examples

7.1 Shells of revolution. The example is used as benchmark demonstrating the principle capabilities of shape optimization in a qualitative way. It also shows the performance of the SQP and MMA solution schemes. The objective is either weight or strain energy. Starting from a spherical shell of uniform thickness (radius 10.0 m) with an apex hole of radius 2.5m the optimal shape is obtained for four different load cases: concentrated ring load c = 255 kN/m on the upper free edge, dead load (γ = 78.5 kN/m^3), uniform vertical load of 0.75 kN/m^2 (snow)

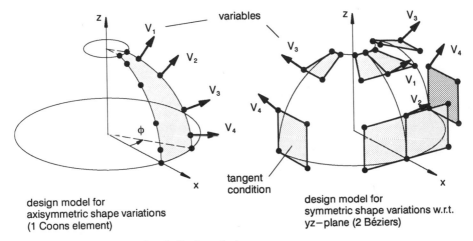

Fig. 14. *Design elements for shell of revolution.*

and a uniform horizontal load of 0.72 kN/m² with pressure on one side and suction on the other side roughly simulating wind. Either a fixed hinged or a horizontal roller support is assumed. The material properties of steel are chosen: $E = 2.1 \cdot 10^5$ N/mm²; $\upsilon = 0.3$. Stress constraints with a v. Mises stress limit of 160 N/mm² are considered.

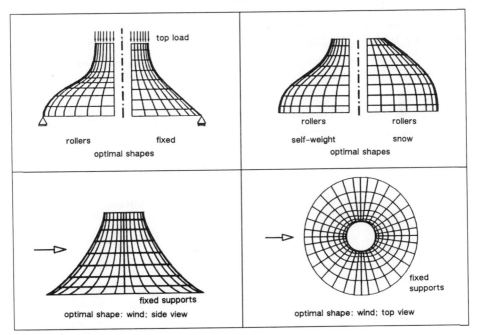

Fig. 15. *Shape optimization of shell of revolution.*

load	support	objective	improve-ment [%]	optimizer	no. of iterations
concentrated load	roller	weight	86	MMA	10
concentrated load	fixed	weight	75	MMA	12
dead load	roller	strain energy	3.5 *	SQP	3 *
snow	roller	weight	63	MMA	15
"wind"	fixed	strain energy	88	SQP	9

Table 2. *Parametric study for shell of revolution; (* pre–optimized initial design).*

The surface can be defined by one single Coons element described in cylindrical coordinates using a 6–node B–spline and a 2–node Lagrange side in meridional and hoop direction, respectively (Fig. 14, left). Since the optimization for the horizontal load leads to a non–axisymmetric shell two Bézier elements are applied for one half of the structure (Fig. 14, right). In both cases only 4 geometrical design variables V_i are introduced. In case of weight minimization the uniform thickness is taken as extra variable.

Figure 15 shows the optimal shapes for different loads and boundary conditions. Except for the "wind load" the final shape allows the shell to carry the load by almost pure membrane action. This holds for weight (concentrated and snow load) as well as strain energy (dead load) minimization. Table 2 summarizes the main results of the parametric study.

7.2 Fly–wheel. The fly–wheel shown in Fig. 16 is designed for optimal strength using shape and sizing variables simultaneously. The whole structure consists of parts with different stress states. The flange acts principally as a ring in tension with combined bending in two directions whereas the inner part of the wheel is a membrane. Any geometric variation of the structure causes a highly nonuniform stress state. During the design procedure the rotational inertia of the wheel has to be preserved by an equality constraint and the optimization problem can be stated as:

$$\text{minimize:} \quad f_s(x) = \int (\sigma_v - \sigma_{all})^2 \, dv$$

$$\text{subject to:} \quad h_C = \frac{C}{C^0} - 1 = 0 \quad ; \quad C = \int \varrho \, r^2 \, dv$$

$$g_\sigma = \frac{\sigma_V}{\sigma_{all.}} - 1 \leq 0 \quad ; \quad \sigma_{all} = 1.2 \times 10^5 \frac{kN}{m^2}$$

Shape optimization with program CARAT

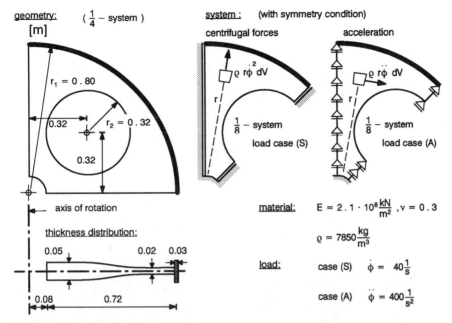

Fig. 16. *Fly–wheel: problem statement.*

The geometry of the structure as well as its variations are supposed to be multiply symmetrical (Fig. 16). Two different load cases are considered which take centrifugal forces (S) and acceleration (A) into account. Both load cases can be analysed for 1/8 of the whole structure by modifying the boundary conditions. The stress state due to combined loading is obtained by symmetric and anti–symmetric superposition. The problem functions have to be assigned to the resulting stress state.

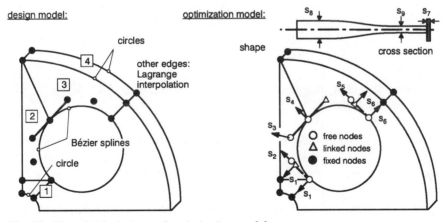

Fig. 17. *Fly–wheel: design and optimization model.*

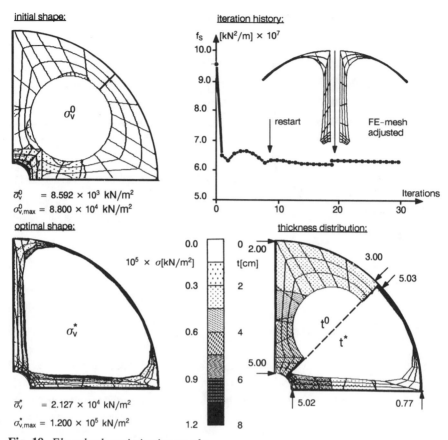

Fig. 18. *Fly–wheel: optimization results.*

The geometrical layout is described by a few optimization variables. 14 design nodes and 6 Coons patches are used (Fig. 17) where Bézier splines describe the inner holes and circles the outer and inner contours. Adjacent Bézier splines are connected by tangent elements. The linking rules "prescribed move direction" and "change of basis" reduce the number of shape variables to six (Fig. 17). In addition three sizing variables are introduced which describe constant flange thickness and cubic thickness distribution over the membrane part. For this reason shape variations cause simultaneous variations of the thickness distribution.

The iteration history (Fig. 18) reflects the high interaction of sizing and shape variables which is influenced by the equality constraint. NLPQL was used as optimizer. After twenty iteration steps the interactive control shows a distortion of the finite element mesh (8–node reduced integrated shell elements). Without changing the overall optimization problem the element mesh is adjusted by additional pre–processor meshing commands. The optimal structure satisfies the stress constraints at three different locations: at the axis of rotation, the spoke, and the web/flange zone. The result is mainly controlled by the equality constraint which causes the flange to become as thick as possible by a low stress level.

7.3 Shape Optimization of a Bell. The major design aspect for a bell is to preserve high tonal quality. For this reason the tuning of the basic frequencies of the bell is introduced as objective of the optimization problem without any other constraints (Kimmich et al. 1992). The frequency requirements of a minor– and a major– third bell is very much influenced by the number n of goal frequencies a_{i0} and the individual weighting factors w_i used in the objective f_T:

$$f_T = \sum_{i=1}^{n} \frac{(a_i - a_{i0})^2}{a_{i0}^2} w_i \qquad (41)$$

This tuning problem is a multicriterion optimization problem. As optimization strategy a SQP–method is used. Since all frequency requirements are met exactly at the optimal solution, the individual weighting factors w_i are only important for the convergence of the algorithm.

The variables for the optimization model are selected with special care to obtain a well posed optimization problem. As shown in Fig. 19 eight sizing and ten shape values in the design model are used as relevant optimization variables. Some restrictions for design coordinates and nodal thickness values are introduced to obtain a useable optimization result. The number of variables can be further reduced by using linking schemes for sizing and shape variables, like the vertical shape coordinates in Fig. 19, which all are linked to the variable s_1.

From the above mentioned 17 relevant optimization variables 15 are used as independent variables to improve the minor–third bell. Five of them are sizing variables. The resulting shape (Fig. 20) shows only slight modifications of the initial shape.

To obtain the major–third bell, the height of the bell is introduced as additional variable. The shape of this bell has an increased height and a moderate bump at half the height of the bell, also described by Schoofs et al. (1985).

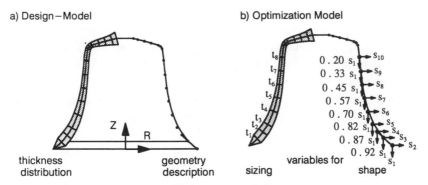

Fig. 19. *Definition of the optimization model of the bell.*

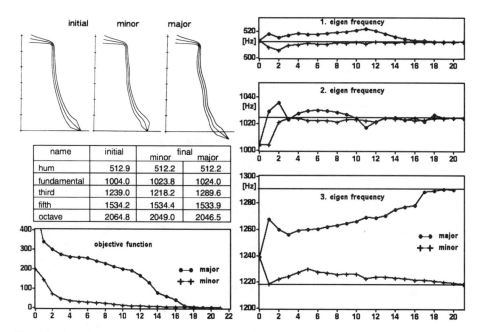

Fig. 20. *Optimization results and iteration history of minor– and major–third bell.*

In both optimization procedures the objective function becomes zero within a given tolerance bound. This means that the frequency requirements are fulfilled exactly (see final frequencies in table of Fig. 20). The iteration history of the objective function and the lower three eigenfrequencies are shown in Fig. 20 for both, the minor– and major–third bell. The difference is pointed out by the third eigenfrequency, which is getting a higher value in optimization of the major–third bell.

7.4 Topology Optimization. This example shows a further capability of CARAT which deals with the optimization of material (i.e. thickness) distribution to generate optimal topologies (Fig. 21). The procedure is a variant of the homogenization method by Bendsøe and Kikuchi (1988). The aim is to minimize the strain energy of a structure of given geometry and total material volume where the thickness distribution is variable. Homogenous material is assumed, the objective function is strain energy, the variables are the finite element thicknesses which are assumed to be constant within each element (refer to Table 1). Since the number of optimization variables is usually large, MMA or a modified "fully stressed design" update which is able to consider the volume equality constraint are used as optimizers.

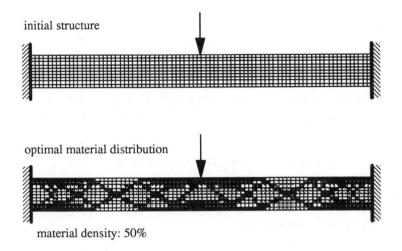

Fig. 21. *Optimal topoloy of a clamped beam generated by FSD method.*

8. Conclusions

Some of the important features of CARAT concerning structural optimization are outlined. The program is dedicated to research and is designed as a flexible and easy to expand but modular computational tool. The program integrates design, structural, and optimization models into a functional unity. It is developed on the basis of a common in–core data structure which serves with high efficiency and theoretical as well as algorithmic generality. The developments are mainly directed towards shape optimization of shells which is reflected by special features of the design model (e.g. Bézier– and B–spline approximations, geometrical continuity) and high requirements on the optimization algorithms. In that context a consistent extension of the method of moving asymptotes is presented. Very good experiences with SQP methods are stated. Some examples show the variety of possible applications including a possible approach to topology optimization. Current research is concerned with optimization of structures with nonlinear behavior and simultaneous mesh adaptation and shape optimization.

Acknowledgements

This work is part of the research project SFB 230 "Natural Structures – Light Weight Structures in Architecture and Nature" supported by the German Research Foundation (DFG) at the University of Stuttgart. The support is gratefully acknowledged.

References

Atrek E., Gallagher R. H., Ragsdell K. M. and Zienkiewicz O. C. (1984, eds.), New Directions in Optimum structural Design. Wiley, Chichester, New York.

Bendsøe M. P. and Kikuchi N. (1988), Generating optimal topologies in structural design using a homogenization method. Computer Methods in Applied Mechanics and Engineering, vol. 71, 197–224.

Bennett J. A. and Botkin M. E. (1986, eds.), The Optimum Shape – Automated Structural Design. Plenum Press, New York, London.

Bletzinger K.-U. (1990), Formoptimierung von Flächentragwerken. Ph. D. Dissertation, Institut für Baustatik, Universität Stuttgart.

Bletzinger K.-U., Kimmich S. and Ramm E. (1992a), Efficient modeling in shape optimal design. Computing Systems in Engineering, vol. 2, 483–495.

Bletzinger K.-U. and Ramm E. (1992b), Form finding of shells by structural optimization. To appear in Engineering with Computers.

Bletzinger K.-U. (1992c), Extended method of moving asymptotes based on second order informations. Accepted for publication in Structural Optimization.

Böhm W., Farin G. and Kahmann J. (1984), A survey of curve and surface methods in CAGD. Computer Aided Geometric Design, vol. 1, 1–60.

Braibant V. and Fleury, C. (1986), Shape optimal design and CAD oriented formulation. Engineering with Computers, vol. 1, 193–204.

Eschenauer H. A. and Thierauf G. (1989a, eds.), Discretization Methods and Structural Optimization – Procedures and Applications. Proceedings of the GAMM–Seminar of the same name, Oct. 5–7, 1988, Siegen, Lecture Notes in Engineering, Springer, Heidelberg, Berlin.

Eschenauer H., Post U. and Bremicker M. (1989b), Einsatz der Optimierungsprozedur SAPOP zur Auslegung von Bauteilkomponenten. Bauingenieur, vol. 63, 515–526.

Faux I. D. and Pratt M. J. (1979), Computational Geometry for Design and Manufacture. Ellis Horwood Publishers, Chichester.

Fleury C. (1986), Structural optimization – a new dual method using mixed variables. International Journal for Numerical Methods in Engineering, vol. 23, 409–428.

Gallagher R.H. (1973), Fully stressed design. In: Optimum Structural Design – Theory and Applications (edited by R. H. Gallagher and O. C. Zienkiewicz), J. Wiley, London, New York, 1973.

Haftka R. T. and Grandhi R. V. (1986), Structural shape optimization – a survey. Computer Methods in Applied Mechanics and Engineering, vol. 57, 91–106.

Haftka R. T. and Adelman H. M. (1989), Recent developments in structural sensitivity analysis. Structural Optimization, vol. 1, 137–151.

Haftka R. T., Gürdal Z. and Kamat M. P. (1990), Elements of Structural Optimization. 2nd edition, Kluwer Academic Publishers, Dordrecht.

Kimmich S. (1990), Strukturoptimierung und Sensibilitätsanalyse mit finiten Elementen. Ph.D. Dissertation, Institut für Baustatik, Universität Stuttgart.

Kimmich S., Reitinger R. and Ramm E. (1992), Integration of different numerical techniques in shape optimization. To appear in Structural Optimization, vol. 4, 1992.

Mota Soares C. A. (1987, ed.), Computer Aided Optimal Design – Structural and Mechanical Systems. NATO–ASI Series F: Computer and System Sciences, vol. 27, Springer, Berlin, Heidelberg.

Philippi U. and Nachtigall W. (1991), Constructional morphology of sea urchin tests. In: proceedings of the 2nd International Symposium on *Natural Structures – Principles, Strategies, and Models in Architecture and Nature*, Stuttgart, Oct. 1 to 4 1991, Mitteilungen des SFB 230, Heft 6, Sonderforschungsbereich 230, Stuttgart, 183–191.

Prasad B. (1983), Explicit constraint approximation forms in structural optimization, part 1: analyses and projections. Computer Methods in Applied Mechanics and Engineering, vol. 40, 1–26.

Ramm E., Bletzinger K.–U. and Kimmich, S. (1991a), Strategies in shape optimization of free form shells. in: Nonlinear Computational Mechanics – State of the Art (edited by P. Wriggers, W. Wagner),, Springer, Heidelberg, Berlin, 163–192.

Ramm E. and Mehlhorn G. (1991b), On shape finding methods and ultimate load analyses of reinforced concrete shells. Engineering Structures, vol. 13, 178–198.

Ramm E. and Reitinger R. (1992), Force follows form in shell design. Presented at: IASS–CSCE International Congress on "Innovative Large Span Structures", 13–17 July 1992, Toronto, Canada.

Rozvany G. (1991, ed.), Optimization of large structural systems. Proceedings of the NATO/DFG conference of the same name, vol. 1 to 3, Berchtesgaden, 23. Sept. – 4. Oct. 1991.

Schittkowski K. (1981), The nonlinear programming method of Wilson, Han and Powell with an augmented Lagrangian type line search function. Numerische Mathematik, vol. 38, 83–114.

Schmit L. A. (1981), Structural synthesis – its genesis and development. AIAA Journal, vol. 19, 1249–1263.

Schoofs B., van Asperen F., Maas P. and Lehr A. (1985), Experimental design theory and structural optimization design for a major–third church bell. CAMP '85, C 4.2.

Schwefel H. P. (1981), Numerical Optimization of Computer Models. J. Wiley & sons, Chichester.

Starnes J. H. and Haftka R. T. (1979), Preliminary design of composite wings for buckling, strength, and displacements constraints. Journal of Aircraft, vol. 16, 564–570.

Svanberg K. (1987), The method of moving asymptotes – a new method for structural optimization. International Journal for Numerical Methods in Engineering, vol. 24, 359–373.

Vanderplaats G. N. (1982), Structural optimization – past, present and future. AIAA–Journal, vol. 20, 992–1000.

Vanderplaats G. N. (1984), Numerical Optimization Techniques for Engineering Design: With Applications. McGraw–Hill, New York.

Authors' address:

Dr.–Ing. Kai–Uwe Bletzinger,
Dipl.–Ing. Reiner Reitinger,
Dr.–Ing. Stefan Kimmich,
Prof. Dr.–Ing. Ekkehard Ramm

Institut für Baustatik
Pfaffenwaldring 7
D–7000 Stuttgart 80

DYNOPT

A Program System For Structural Optimization Weight Minimum Design With Respect To Various Constraints

Detlev W. Mathias and Herbert Röhrle

Abstract. This paper describes the structural optimization program DYNOPT in its present state, all available types of finite elements and all types of constraints implemented until now. Furthermore the basis of analysis and optimization as well are discussed. Finally the optimization of a wing box is presented and the obtained results are discussed.

1. Introduction

The development of the program system DYNOPT in its present form started in the year 1975. Since then it has been improved and extended continuously until now. The main purpose of the development of this program system was the need of designing mainly aircraft structures such as airframes and their components to minimum weight. Actually the minimum weight design of aircraft structures is of great importance with respect to

- range
- payload
- number of passengers
- manoeuvrability
- maintainability
- production costs
- etc.

Having all this in mind the above mentioned items lead to the need of having a reliable and easily to use optimization procedure.

As the structures or structural components of present airframes are pretty complex there is a whole lot of sizes to be varied in order to obtain a reasonable design which also satisfies predefined constraints. In the case of anisotropic structures the orientation of the anisotropic axes is an additional type of design variables with a great potential for higher benefits in structural weight (see Röhrle H., Mathias D. W. 1986).

The need to design complex structures for minimum weight considering different types of behaviour and side constraints as well directly leads to the demand for optimization techniques.

Due to this fact Dornier is involved in the development and in the application of finite element techniques and optimization procedures as well since many years.

As at the beginning of the eighties could be seen the performance of the computers considerably increased and the computation costs sharply decreased. So, not having in mind computational effort, reliable and highly accurate methods and algorithms could be selected for solving the optimization problems.

Thus the only conclusion could be to consider mathematical programming methods for application in the program system DYNOPT.

From the very beginning in 1975 the program DYNOPT is based on finite element analysis and on sequential linear programming (SLP). The combination of these two methods has proved highly adaptable to structural design problems, and a lot of advantages results from this procedure some of which are listed below.

- mathematical formulation of the optimization problem

- adaptability to a lot of different types of constraints

- reliability and accuracy of the iterative procedure

- excellent convergence behaviour

- the selected methods support an easy extension of the program features, i.e.

    · implementation of additional finite elements
    · consideration of additional constraints
    · different types of design variables

The program system DYNOPT has a modular structure. This results in easy extension work for further features.

DYNOPT is part of the Dornier in-house FEM-program called COSA (COmputational Structural Analysis) although it is a stand-alone program. It has special postprocessing routines due to the iterative process and is furthermore linked to the pre- and postprocessing programs of COSA.

The programming language used in FORTRAN 77. The crucial algorithms such as factorization of the stiffness matrix, determination of the eigenvalues, flutterspeed calculation, sensitivity analysis etc. are running double precision. Installations are done for IBM-3090, CONVEX-C2 and SUN sparcstation at Dornier.

Until now DYNOPT has been used at Dornier company for components of Alpha Jet, EFA, Do 328 and Airbus A 330/340. On the other hand not only aircraft structures may be optimized using DYNOPT but also arbitrary structures or components.

DYNOPT runs are fully automated. In order to increase the applicability of the whole program even for non-specialists the number of parameters to control the individual run has been confined to a minimum.

All input data are checked extensively. Error codes and hints are provided to inform the user about mistakes and the way to correct them.

2. Optimization Model

The goal is to start a structural optimization from an initial design with estimated values for the design variables and to get a minimum weight design satisfying all constraints. The procedure to be used is an iterative process.

The optimization problem consists of an objective function i.e. the minimum weight function and a set of constraints as well

- strain constraints

- displacement and distorsion constraints

- stress constraints

- buckling constraints

- frequency constraints

- aeroelastic efficiency constraints
- flutterspeed constraint

have to be satisfied while the objective function approaches its optimum value.

In the following text the optimization problem is represented by a minimum weight function
$$W = W(x_1, x_2, x_3, \ldots, x_m) = Min. \tag{1}$$
and a set of displacement and failure constraints
$$U = U(x_1, x_2, x_3, \ldots, x_m) \leq U_{adm} \tag{2}$$
$$RF = RF(x_1, x_2, x_3, \ldots, x_m) \geq RF_{adm} \tag{3}$$

The objective function W as well as the set of constraints U and RF are functions of the design variables x which are sizes in general and angles β of the anisotropic axes in the case of compound structures.
$$\{x\}^T = \{t_1, t_2, \ldots, t_j, \beta_1, \beta_2, \ldots, \beta_k\} \ , \quad j + k = m \tag{4}$$
It is assumed that the topology and the shape of the structures to be optimized remain unchanged in the optimization process. So at present the design variables x of an optimization problem may be

- cross-sectional areas of bars
- thicknesses of membranes

In case that anisotropic structures have to be optimized the variables are

- thicknesses of plies or layers
- thicknesses of the compounds
- angles of the anisotropic axes

In addition to the behaviour constraints (2) and (3) a set of side constraints may be formulated. These side constraints are the minimum gauges of the design variables, that in fact means they are production related constraints.
$$\{x\} \geq \{x\}_{min} \tag{5}$$
The design variables x may be linked to groups. In this way for instance it is possible to design weight optimal structures keeping predefined areas at a uniform thickness. But not only uniform thickness is possible but also a factor-controled increase or decrease of the element thicknesses is permitted. In case of moderate element linking the penalty of structural weight will be negligible and the improvement of the design with respect to practicability will be substantial.

3. Structural Analysis

The program system DYNOPT is a stand-alone program. However a lot of components and modules is taken from the Dornier in-house FEM-programs COSA-DEMEL for statical analysis and COSA-DYNAME for dynamical analysis.

Especially the finite elements are taken from the COSA. There are

- bars (constant or linear stress)
- beams (I-beam and box cross-section)
- isotropic triangular membranes with constant stress
- isotropic triangular membranes with linear stress distribution
- isotropic quadrilateral membranes with linear stress distribution
- anisotropic triangular membranes with linear stress distribution
- anisotropic quadrilateral membranes with linear stress distribution
- isotropic triangular shells
- anisotropic triangular shells

Results of the analysis are

- displacements
- slopes
- strains
- stresses (normal and shear)
- v. Mises equivalent stresses for the isotropic membranes and the isotropic shells
- reserve factors for the anisotropic membranes and the anisotropic shells
- buckling factors (for local and global buckling)
- natural frequencies
- aeroelastic efficiency
- flutterspeed

The above mentioned results of the analysis can also be limited to upper or lower boundary values respectively, i.e. they can be made constraints.

All constraints or any arbitrary subset thereof are treated the same way and simultaneously during the optimization process. Thus it is guaranteed, that the obtained optimum is a real optimum.

For each improved design the set of constraints has to be calculated again. In the DYNOPT program of course an active set strategy is used to confine the number of constraints to be calculated to the active ones. These active constraints are the decisive ones that have by far the biggest impact on the determination of the design variables.

However, at least one of each type of the constraints is considered to avoid a flip-flop caused by activating and deactivating.

Basically the system of static equations as it is given by

$$[K][U] = [F] \tag{6}$$

is solved by decomposition of the stiffness matrix $[K]$ and forward - backward - substitution of the load vectors in the matrix $[F]$. At the moment up to 48 loadcases can be considered in one run. The decomposed stiffness matrix is stored for later use.

The reserve factor RF calculated for the compound elements is determined according to a modified Tsai/Hill failure hypothesis. It is calculated from the three in-plane stresses σ_1, σ_2 and τ_{12} related to the corresponding ultimate stresses (index u).

For each of the directions 1 and 2 an ultimate tension stress as well as an ultimate compression stress is necessary. They will be selected dependent on the sign of the normal stresses σ_1 and σ_2.

$$\frac{\sigma_1^2}{\sigma_{1u}^2} + \frac{\sigma_2^2}{\sigma_{2u}^2} + \frac{\tau_{12}^2}{\tau_{12u}^2} = H \tag{7}$$

resulting in

$$RF = \frac{1}{\sqrt{H}} \tag{8}$$

The minimum reserve factor is determined for each layer of the anisotropic elements. Values less than 1.0 are pointing out failure of the concerned layer. Reserve factors exceeding 1.0 indicate that no failure will occur. However for the reserve factor constraints the limit values may be greater than 1.0 in order to consider arbitrary safety factors.

In the case of buckling analysis the geometrical stiffness matrix $[K_g]$ is determined and the following eigenvalue problem will be solved using either Jacobi or preferably a subspace method.

$$([K] + \lambda[K_g])\{q\})\{0\} \tag{9}$$

The matrices $[K]$ and $[K_g]$ as well normally are reduced matrices after a statical condensation (see Hornung G. et al. 1990).

Condensation either is applied to the stiffness matrix $[K]$ and the matrix of inertia $[M]$ in the case that the eigenfrequencies and the eigenmodes have to be determined.

The eigenvalue problem

$$([K] - \lambda[M])\{q\} = \{0\} \tag{10}$$

is normally solved using the Jacobi or a subspace method depending on the number of natural frequencies to be determined. By the way the mass matrix $[M]$ is kinematically equivalent to the stiffness matrix $[K]$.

If aeroelastic efficiency analysis of a fin or wing with or without control surfaces is selected static condensation to just the degrees of freedom perpendicular to the mid-plane is necessary. The following system of linear equations has to be solved

$$([K] - [K_A])\{U\} = \{PA\}^s \tag{11}$$

As the 'aerodynamic stiffness matrix' $[K_A]$ is unsymmetric a Cholesky-decomposition in the basic form is impossible. In the DYNOPT program the solution of (11) currently is obtained by using the Gauss algorithm.

Also for the flutter analysis the same condensation as for the efficiency analysis is performed. For the remaining primary degrees of freedom the following flutter equation has to be solved

$$([K] - \lambda[M] - 0.5pv^2[A])\{q\} = \{0\} \tag{12}$$

In (12) $[A]$ is the matrix of aerodynamic influence coefficients dependent on the Mach number Ma and the reduced frequencies k.

4. Sensitivity Analysis

The behaviour constraints (2) and (3) of the optimization problem as it is described by the equations (1), (2), (3) and (5) generally are nonlinear. To obtain a solution this nonlinear optimization problem will be substituted by a sequence of linearized subproblems. Taylor series expansion of the objective function (1) and the behaviour constraints (2) and (3) as well neglecting higher order terms results in the linearized subproblem

$$\{\nabla W\}^T \{\Delta x\}^{\nu+1} = \Delta W = Min. \tag{13}$$
$$[\nabla U^{(\nu)}]^T \{\Delta x\}^{\nu+1} \leq \{U\}_{adm} - \{U^{(\nu)}\}_{act} \tag{14}$$
$$[\nabla RF^{(\nu)}]^T \{\Delta x\}^{(\nu+1)} \geq \{RF\}_{adm} - \{RF^{(\nu)}\}_{act} \tag{15}$$

The inequalities (14) and (15) represent the whole lot of behaviour constraints as listed in chapter 3. Additional constraints have to be considered for the design variables. These are the side constraints.

$$\{\Delta x\}^{(\nu+1)} \geq \{x\}_{min} - \{x^{(\nu)}\}_{act} \qquad (16)$$

The indices (ν) and $(\nu + 1)$ respectively in equations (13) to (16) mark the iteration step.

The determination of the gradients of the behaviour constraints is done by deriving them with respect to the design variables. For all constraints the analytical way of calculation has been selected. This can be done as none of the implemented FEM-routines is a 'black box' and each kind of information thereof is available for further use in special gradient programs.

Many of the constraints described at the beginning of chapter 3 are of that type that needs the gradients of the displacements. These can be obtained by differentiation of the basic statical equations (6) with respect to the design variables assuming that the loads $[F]$ are not dependent on the design variables.

$$[\nabla U] = -[K]^{-1}[\nabla K][U] \qquad (17)$$

The displacement gradients (17) are part of the displacement constraints (14). They will either be used to determine the reserve factor gradients $[\nabla RF]$, and the stress- and strain gradients $[\nabla \sigma]$ and $[\nabla \epsilon]$. The reserve factors RF are functions of the stresses as is pointed out in (7) and (8) and these on the other hand are functions of the design variables.

The equations

$$[\nabla \sigma] = [S][T_\sigma]^T [\nabla \epsilon] \qquad (18)$$

and

$$[\nabla \epsilon] = [\epsilon_u][T][\nabla U] \qquad (19)$$

are part of the reserve factor gradients which will be taken to set up the failure constraints for compound structures (see also Mathias D.W., et al. 1988).

All gradients are calculated analytically. This results in a good and reliable convergence without numerical problems during the whole optimization process.

After setting up the complete optimization problem consisting of

- the objective function

- the behaviour constraints as described above

- the side constraints

the improvement of the structural design must be determined. This will be done by means of the SLP.

5. Optimization Algorithms

Two optimization algorithms have been implemented into the program DYNOPT

– the fully stressed design

– the sequential linear programming

The fully stressed design has just been used at the very beginning of the development of DYNOPT for problems with stress constraints only.

Since then the sequential linear programming has been applied to optimization problems. Good results can be obtained without having convergence problems. The Simplex algorithm is used in the DYNOPT program to determine the changes of the design variables.

The range of the design variable changes is limited to lower and upper bounds in order to improve the convergence. For a wide range oszillations may occur due to the fact that the constraints are nonlinear. On the other hand the calculation effort may be increased unnecessarily by a narrow range for the change of the design variables.

In DYNOPT the initial stepsize is part of the input. As structural improvement proceeds the move limits decrease. There is a lower limit for the stepsize either.

$$\{\Delta x\}_{ll} \leq \{\Delta x\} \leq \{\Delta x\}_{ul} \tag{20}$$

In order to speed up the optimization process an active set strategy is used. This means in particular that just the constraints close to their limits and violated constraints are considered in the Simplex algorithm, i.e. those having the biggest impact on the solution vector. The activity of each constraint is checked for each iteration step. Conditions for activity are as follows in the case of displacement constraints

$$\begin{aligned} \{U\}_{ll} &\leq \{U\} \leq \{U\}_{adm} \quad \text{in the feasible region} \\ \{U\}_{adm} &\leq \{U\} \quad \text{in the unfeasible region} \end{aligned} \tag{21}$$

It is taken care that at least one of each type of constraints is active, i.e. the most critical.

The improved structural design is obtained by superposition of the current design and the changes of the design variables.

$$\{X\}^{(\nu+1)} = \{X\}^{(\nu)} + \{\Delta x\}^{(\nu+1)} \qquad (22)$$

The iteration process will be stopped if the convergence criteria are satisfied. These are in particular

- the obtained design is a feasible design
- the values of the objective function change less than a limit value
- the design variables change less than a limit value

Are one or more of the convergence criteria violated another optimization step is done. This procedure will continue as long as necessary to satisfy the criteria or until a maximum number of iteration steps is done.

A rough flow chart of the optimization program DYNOPT is shown in Fig. 1. The blocks of Finite Element Analysis and Optimization are especially marked.

6. Optimization of a Wing Box

An anisotropic wing box of a swept wing has been selected for an optimization considering displacement and reserve factor constraints. The design variables of the structure are the sizes of the finite elements and the orientation of the anisotropic axes in the upper and lower cover as well.

The wing box (Fig. 2) is a clamped structure with three spars and four ribs carrying three different air pressure distributions as statical load cases.

The z-displacements of the wing tip (see Fig. 3) and the reserve factors in all composite elements as shown in Fig. 4 are limited.

The type of element used in this optimization is a quadrilateral compound element with linear stress distribution along the edges. For each of the meshes the thickness of the $0°-$, $\pm 45°-$ and $90°-$layers will be variable.

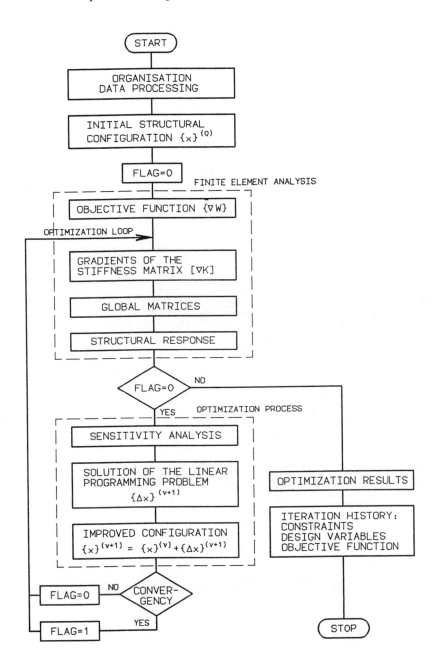

FIG. 1 : FLOW CHART OF THE OPTIMIZATION PROGRAM DYNOPT

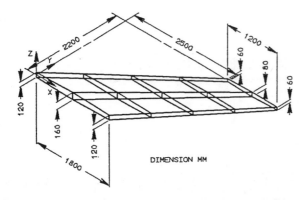

FIG. 2: WING BOX OF A SWEPT WING

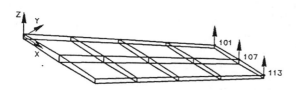

FIG. 3: FE-MODEL OF THE WING BOX WITH THE CONSTRAINED Z-DISPLACEMENTS

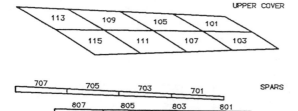

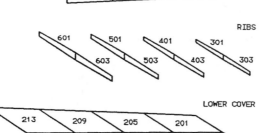

FIG. 4: FE-MODEL OF THE WING BOX WITH ANISOTROPIC QUADRILATERAL ELEMEN

Some characteristic figures of this simple structural optimization model of the wing box are

- 89 nodal points
- 228 degrees of freedom
- 92 anisotropic quadrilateral membranes
- 83 design variables (including 2 variable angles β)
- 261 constraints alltogether

In Fig. 5 the changes of the orientation of the anisotropic axes can be seen for the upper cover (SYST. 1) and the lower cover (SYST. 2) as well plotted against the iterations. As could be expected both angles β are equal due to the structure and the applied load cases just after few iterations. The angles β are measured against the global x-axis. Decreasing angles β mean rearrangement of the compound axes along the loadpath.

Active constraints throughout the whole optimization process have been the displacement in dof 101 Z (leading edge of the wing tip) and the slope of the tip rib as can be seen from Fig. 6.

The reserve factor constraints have been active for the elements along the line between the leading edge of the wing tip and the trailing edge of the wing at the support. The initial design of the wing is infeasible with respect to displacement as well as reserve factor constraints.

The iteration history of selected elements is presented in Fig. 7 to 11 by means of reserve factor plots and plots of the sizes.

Noticeable in the mass plot (Fig. 12) is the peak in the first iteration step although the initial mass already is considerably exceeding the final mass. This is caused by a general stiffness redistribution. The distribution of the final thicknesses (Fig. 13) has come out like it has been expected. The stiffness concentration is along a line between the leading edge of the wing tip and the trailing edge of the wing support following the main load path.

7. Summary

The optimization program DYNOPT in its present form has been developed at Dornier Luftfahrt GmbH since 1975. It represents a useful and powerful tool to minimize the weight of arbitrary structures under consideration of various types of constraints. Design variables are the cross-sectional areas, thicknesses of membranes and shells and the orientations of the anisotropic axes. Design variable linking and mini-

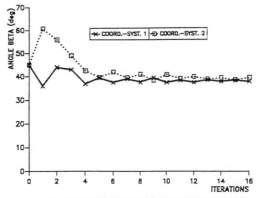

FIG. 5 : VARIATION OF THE ANISOTROPIC AXES

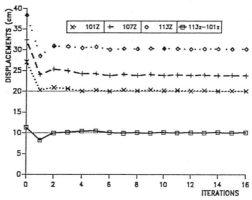

FIG. 6 : CONSTRAINED Z-DISPLACEMENTS

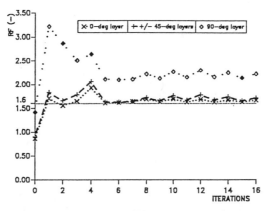

FIG. 7 : RESERVEFACTOR OF COVER ELEMENT 115

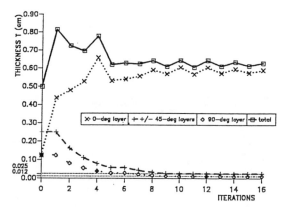

FIG. 8 : THICKNESS OF THE LAYERS OF COVER ELEMENT 115

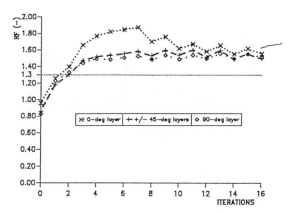

FIG. 9 : RESERVEFACTOR OF COVER ELEMENT 215

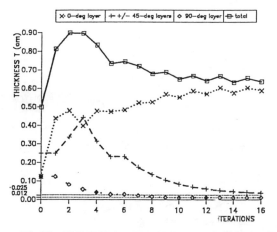

FIG. 10 : THICKNESS OF THE LAYERS OF COVER ELEMENT 215

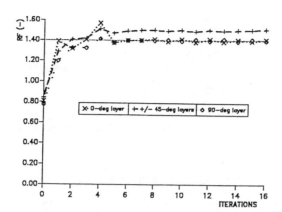

FIG. 11 : RESERVEFACTOR OF SPAR ELEMENT 907

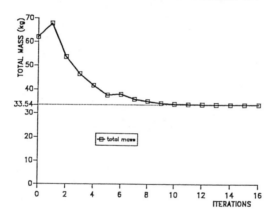

FIG. 12 : ITERATION HISTORY OF THE STRUCTURAL MASS

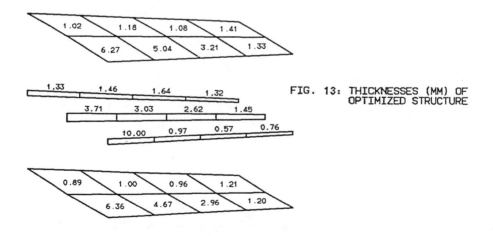

FIG. 13: THICKNESSES (MM) OF OPTIMIZED STRUCTURE

mum gauges are among the features for manufacturing oriented design. Measures to reduce calculation effort are implemented, such as active set strategy. DYNOPT is a stand-alone program which is completely compatible to the Dornier in-house FEM program COSA. Pre- and postprocessing will be done using COSA routines. Iteration history tables and diagrams are part of the final output.

The basis of DYNOPT are the finite element techniques and the sequential linear programming. Numerical stability and convergence of the iterative procedure are good.

The program DYNOPT has been successfully applied to various optimization problems in the design of the Alpha Jet, EFA fin, Do 328 and Airbus components.

References

RÖHRLE, H., MATHIAS, D.W. (1986): *Möglichkeiten und Methoden der Strukturoptimierung,* paper presented at the DGLR-Jahrestagung 1986, Munic, (Germany), Jahrbuch 1986, Volume II, Pages 731–739.

HORNUNG, G. ET AL. (1988): *Optimzation of Wing Type Structures Including Stress and Buckling Constraints,* paper presented at the Third Air Force/NASA Symposium on Recent Advances in Multidisciplinary Analysis and Optimization, September 1990, San Francisco (USA).

MATHIAS D.W., ET AL. (1988): *Variation of Anisotropic Axes Due to Multiple Constraints in Structural Optimization,* presented at the 16th ICAS-Congress, 1988, Jerusalem, Israel, proceedings of 16th ICAS-Congress, 1988, Volume II, Pages 1498–1504.

Author's address:

Dipl.-Ing. Detlev W. Mathias
Prof. Dr.-Ing. Herbert Röhrle
DORNIER Luftfahrt GmbH
P.O. Box 1303
D-7990 Friedrichshafen 1
Germany

$MBB - Lagrange$: A Computer Aided Structural Design System

Rainer Zotemantel

Abstract. $MBB - Lagrange$ is a computer aided structural design system which allows the optimization of structural systems. It is based on the finite element technic and mathematical programming. The optimization model is characterized by the design variables and many different restrictions. As design variables are available: element thicknesses, cross sections, concentrated masses, and fiber angles. Besides isotropic, orthotropic, and anisotropic applications the calculation of composite structures is a main point. The design can be restricted with respect to statics (e.g. stresses, buckling), dynamics (natural frequencies, dynamic response) and aeroelastics (effiencies, flutter speed). The general aim is to minimize the structure weight subject to the requirements. But often the question is to get a feasible design, and the weight is less important. Some special features round off this optimization tool. The highlights of the program system are simultaneous optimization, a wide range of optimization strategies, and a modular architecture. The application problems - especially space and aircraft structures - show the capacity of $MBB - Lagrange$.

1. Introduction

The decision to develop a computer aided structural design system was based on the knowledge that the structures especially in space and aircraft become more and more complex. To find a feasible design under the condition of limited technical and economic ressources a powerful tool is worthwhile. This tool should support the engineer in design and analysis of complex structures. These structures often have - on the one hand side - a lot of degrees of freedom and - on the other hand side - conflicting constraints. To solve such problems only the use of computer aided tools based on mathematical optimization is a promising way. $MBB - Lagrange$ is a program system which fulfills these requirements. It became one of the earliest optimization codes including some special features with respect to aircraft design. The program system $MBB - Lagrange$ has been under development since 1984 using a team of engineers and mathematicians.

2. Computer Aided Optimal Design

Complex and large projects require the application of modern tools within the process of computer aided engineering (CAE) which comprises big computer program systems for pre-processing, structural analysis, and post-processing. An important part is the structural optimization.

Let us focus two different tasks. One has a structure of very high complexity and conflicting or very strong requirements with the goal to minimize the structural weight. Then an optimization tool can be the only chance to solve this problem. May be, some iteration loops are necessary but you will find a solution if there is one. Especially in space and aircraft design you will find such problems. A low weight means e.g. reduction of energy or more pay load.

The other task concerns structures which will be produced in a large number of pieces. If you can reduce the weight - sometimes a little for one piece - you gain a lot of money in the whole production.

In many cases the question is not: Which is the best design?, but: Which is a better (and feasible) design?

MBB−Lagrange may be an additional tool to make the daily work of engineers easier. To use an optimization tool the engineer should know what an 'optimization model' means.

MBB − Lagrange combines the method of finite element method (FEM) with mathematical programming algorithms. This leads to the so called Three Columns Concept of Eschenauer (1985) which will be discussed later.

One feature of *MBB − Lagrange* is the open architecture and a modular program organization. The general program architecture is shown in fig. 1.

The program system is splitted into two parts: input and design part. They are linked by a database and a control file.

The first part - input - reads and checks the input data. In the main part - design - the optimization process takes place. There is a strict seperation between analysis and optimization. The optimization procedure is organized due to the Three Columns Concept mentioned above which can be seen crearly in fig. 1.

Many interfaces enable the engineer to generate the model and to check the results. Namely there are interfaces to MSC/NASTRAN (input and output), I-DEAS (pre- and postprocessing), and PATRAN (postprocessing, history plots). Fig. 2 shows the CAE-environment of *MBB − Lagrange*.

3. Optimization Model

The structural optimization problem is formulated as the classical nonlinear mathematical programming problem:

$$min f(\mathbf{x}), \quad \mathbf{x} \epsilon R \qquad (1)$$

MBB-Lagrange: a computer aided structural design system

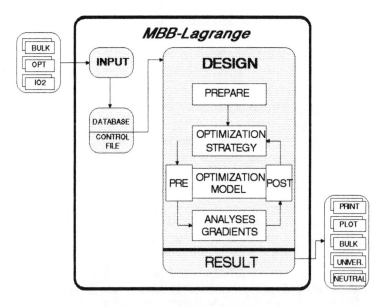

Figure 1: General program architecture

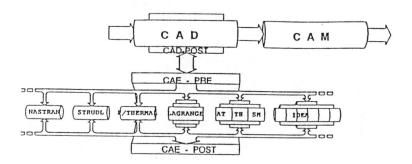

Figure 2: CAE-environment of $MBB - Lagrange$

subject to
$$g_j(x) \geq 0, \quad j = 1, ..., m_g \qquad (2)$$

$$x_{il} \leq x_i \leq x_{iu}, \quad i = 1, ..., n \qquad (3)$$

In the above equations f means the objective function, $\mathbf{x}$ the design variables, x_{il} and x_{iu} the lower and upper bounds, resp., and g_j the constraints.
In $MBB - Lagrange$ the objective function $f(\mathbf{x})$ is

$$minimum \quad weight$$

But it is also possible to take other linear objective functions. Besides one can optimize problems with several objective functions, e.g. weight and stresses. For such multiobjective problems one must use the methods of vector optimization.

The design variables can be devided into four types:

1. sizing variables, i.e.
   - cross sectional areas for trusses and beams
   - wall thicknesses for membrane and shell elements
   - laminate thicknesses for every single layer in composite elements
2. balance masses
3. angles of layers for composite elements

An important part are the constraints, here formulated as inequality constraints

$$g = 1 - \frac{r_{act}}{r_{allow}} \geq 0 \qquad (4)$$

The constraints make the structure feasible and realistic with respect to manufacturing. In principle constraints have to be dealt with simultaneously as they restrict the design space in which the optimized structure will ultimately be found. Using gage constraints preconditions for manufacture are concluded. Which constraint combination is to be applied depends on the physical problem. The following constraints are available in $MBB - Lagrange$:

- displacements
- stresses
- strains
- buckling (critical stresses, wrinkling)
- local compressive stresses
- aeroelastic efficiencies
- flutter speed
- natural frequencies
- dynamic responses
- eigenmodes
- weight
- bounds for the design variables (gages)

The objective function, the design variables, and the constraints describe the optimization model. It is part of the Three Columns Concept which bases on three 'columns':

(i) optimization algorithm

(ii) structural model (structural response and gradients)

(iii) optimization model as the link between (i) and (ii)

Two main aspects of the optimization model should be mentioned here. First, it is reasonable to scale all significant values for numerical reasons. Now the second aspect. In principle each structural variable as cross section area or wall thickness is a design variable. But for some reasons it is not necessary or even not convenient to handle such an optimization model. In large scale problems one can have a large number of structural variables. On the other hand the structure itself or the loading or manufacturing requirements suggest to link structural variables together. This variable linking means that several structural variables make one design variable. From the mathematical point of view this is a transformation: $x(t) = a + \mathbf{A}t$ with the linking matrix $\mathbf{A}$. It is also possible to fix elements which means these values do not change during the optimization process.

To illustrate an optimization problem in an easy way we take only two design variables and two constraints. Fig. 3 demonstrates this case with the solution.

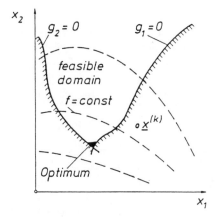

Figure 3: Simple optimization problem

4. Structural Response and Gradients

The structural and sensitivity analysis is based on the finite element method. Modules for the following calculations are included:

- static
- (local) buckling
- natural frequencies
- dynamic responses (frequency, transient, random)
- (stationary) aeroelastic
- flutter

It is possible to treat homogeneous materials with

- isotropic
- orthotropic
- anisotropic

behaviour as well as

- composite materials

The element library contains all important element types:

- truss elements
- beam elements
- membrane elements (triangle, quadriliteral)
- shell elements (triangle, quadriliteral)
- some special elements (e.g. spring elements)

An interesting feature is a special transfer matrix procedure for cylindric shells. Using so called hybrid elements these elements can be assembled together with the remaining finite element stiffness matrix. This mixed formulation allows very efficient analyses in special applications.

The goal is to solve the wellknown equation for example in static analysis

$$\mathbf{K}(\mathbf{x})\mathbf{u}(\mathbf{x}) = \mathbf{p}(\mathbf{x}) \tag{5}$$

with the stiffness matrix $\mathbf{K}$, the displacement vector $\mathbf{u}$, the loading vector $\mathbf{p}$, and the design variables $\mathbf{x}$. After getting the unknown displacements by solving equ. (5) one can calculate the structural response values $\mathbf{r}$, e.g. stresses and strains. Similar equations are available for the other structural analyses mentioned above. From these response values one can obtain the constraint function values:

$$\begin{aligned} \mathbf{g}_j(\mathbf{x}) &= \mathbf{g}_j(\mathbf{r}(\mathbf{x}, \mathbf{u}(\mathbf{x}))) \\ j &= 1, ..., m_g \end{aligned} \tag{6}$$

We have to distinguish active and passive constraints. The violated constraints are always active. Violated means equation (2) is not fulfilled by evaluating equ. (4). The constraints have different influence on the optimization process. To reduce the number of constraints (in a flexible way) we need only the active constraints, that means these constraints lie below a user specified value. The passive constraints are not of interest. Now we come to an important point. To apply mathematical programming we need the gradient information, $\partial f/\partial x$ and $\partial g/\partial x$. Therefore the evaluation of gradients is essential for an optimization code. One reason is the mathematical programming needs this gradient information to lead the optimization process. On the other hand this information gives a measure for the state and convergence of the found design. Because the evaluation of gradients takes up the largest part of computing time during the optimization process the main point is the effective evaluation. There are three possibilities to obtain the gradient, i.e. the sensitivity analysis:

(i) numerical difference formula
(ii) analytical formula
(iii) semi-analytical formula

In $MBB - Lagrange$ all three cases are realized.

(i) The numerical difference formula is the known approximation

$$\frac{\partial g_j}{\partial x_i} \approx \frac{g_j(\mathbf{x} + \epsilon x_i \mathbf{e}_i) - g_j(\mathbf{x})}{\epsilon x_i} \tag{7}$$

$$i = 1, ..., n$$
$$j = 1, ..., m_g$$
$$\mathbf{e}_i : \text{unit vector}$$
$$\epsilon : \text{small real number}$$

If you consider that for every quotient (i.e. design variable) a full analysis is necessary you can imagine this way is very time consuming. That is not applicable for real life problems. The time is increasing with the power of 4 due to the number of design variables.

(ii) The most effective way is to derive analytical formulae for the sensitivity analysis. With equ. (5) and (6) one can write for example in statics

$$\frac{\partial g_j}{\partial x_i} = \frac{\partial g_j}{\partial \mathbf{r}} \frac{\partial \mathbf{r}}{\partial x_i} + \frac{\partial g_j}{\partial \mathbf{r}} \frac{\partial \mathbf{r}}{\partial \mathbf{u}} \mathbf{K}^{-1}(-\frac{\partial \mathbf{K}}{\partial x_i}\mathbf{u} + \frac{\partial \mathbf{p}}{\partial x_i}) \tag{8}$$

The computing time for calculation equ. (8) depends mainly on the second term including $\mathbf{K}^{-1}$. But note that for the solution of the stiffness equation (5) a decomposition of $\mathbf{K}$ has already performed and only a forward-backward substitution is now needed. If the term

$$\mathbf{K}^{-1}(-\frac{\partial \mathbf{K}}{\partial x_i}\mathbf{u} + \frac{\partial \mathbf{p}}{\partial x_i}) \tag{9}$$

is calculated first and pre-multiplied wih the remaining part then it is called the design space or direct method. Otherwise , first calculate

$$\frac{\partial g_j}{\partial \mathbf{r}} \frac{\partial \mathbf{r}}{\partial \mathbf{u}} \mathbf{K}^{-1} \tag{10}$$

then it is called the state space or adjoint variable method.
Sizing problems allow the explicit formulation of the dependence between stiffness matrix $\mathbf{K}$, the load vector $\mathbf{p}$, and the design variables $\mathbf{x}$. So the derivatives $\partial \mathbf{K}/\partial \mathbf{x}$ and $\partial \mathbf{p}/\partial \mathbf{x}$ can be written directly and implemented in the program code. This leads

to a very efficient way of calculating the gradients for the optimization, especially for large scale problems.

The table shows a summary of system equations and derivatives for all structural responses applicable in $MBB - Lagrange$.

STRUCTURAL RESPONSE	SYSTEM EQUATIONS	DERIVATIVES
structural weight	$W = c^T x$	$\partial W/\partial x_i = c_i$
displacements	$K u = p$	$\partial K/\partial x_i \, u + K \, \partial u/\partial x_i = \partial p/\partial x_i$
stresses	$\sigma = S u$	$\partial \sigma/\partial x_i = \partial S/\partial x_i \, u + S \, \partial u/\partial x_i$
strains	$\varepsilon = B u$	$\partial \varepsilon/\partial x_i = \partial B/\partial x_i \, u + B \, \partial u/\partial x_i$
aeroelastic efficiencies	$K u = p + Q u$	$\partial(K-Q)/\partial x_i \, u + (K-Q) \, \partial u/\partial x_i = \partial p/\partial x_i$
buckling	$(K - \lambda K_g) u = 0$	$\partial(K-\lambda K_g)/\partial x_i \, u + (K-\lambda K_g) \, \partial u/\partial x_i = 0$
frequencies	$(K - \omega^2 M) u = 0$	$\partial(K-\omega^2 M)/\partial x_i \, u + (K-\omega^2 M) \, \partial u/\partial x_i = 0$
flutter speed	$(K - \omega^2 M - \lambda Q) u = 0$	$\partial(K-\omega^2 M-\lambda Q)/\partial x_i \, u + (K-\omega^2 M-\lambda Q) \, \partial u/\partial x_i = 0$
transient response	$M \ddot{u} + C \dot{u} + K u = f$	$\partial K/\partial x_i \, u + K \, \partial u/\partial x_i + \partial C/\partial x_i \, \dot{u} + C \, \partial \dot{u}/\partial x_i + \partial M/\partial x_i \, \ddot{u} + M \, \partial \ddot{u}/\partial x_i = \partial f/\partial x_i$

Figure 4: System equations and derivatives of structural responses

(iii) This semi-analytical formula must be used for geometry variables which can be handled in a test version of $MBB - Lagrange$.

At the end of this chapter one important point must mentioned here. For a successful optimization run it is indispensible that the finite element model - which means the idealization of the real problem - is good enough, will say consistent, robust, and reliable, to give the right structural response. On the other hand the optimization is an aid to show errors resp. inconsistencies of the finite element model and is therefore a tool to check the approximation.

5. A Set of Optimization Algorithms

If you remember the Three Columns Concept one important part is the optimization strategy or algorithm. All algorithms approach the optimum by iteration steps:

$$x^{n+1} = x^n + a^n s^n \qquad (11)$$

where

a : $step\ size$
s : $search\ direction$

In the literature many algorithms are known and each of them has advantages and disadvantages. As $MBB - Lagrange$ was developed to solve a wide range of optimization problems with different properties it is worthwhile to offer a variety of optimization strategies. The capabilities of each algorithm differ from one to another so that they cannot be applied to the same physical problem with the same efficiency. In fact an algorithm can fail for a certain problem. Therefore the user must have the possibility to choose out of a range of optimization strategies depending on this particular problem. In this context strategy or algorithm has not the meaning in a strict mathematical sense. The following algorithms are available in $MBB - Lagrange$:

(1) IBF : Inverse Barrier Function, Morris (1982)
(2) MOM : Method Of Multipliers, Schuldt (1975)
(3) SLP : Sequential Linear Programming, Kneppe (1985)
(4) SRM : Stress Rationing Method, Haftka and Kamat (1985)
(5) RQP1 : Recursive Quadratic Programming, Schittkowski (1986/86)
(6) RQP2 : Recursive Quadratic Programming, Powell (1984)
(7) GRG : Generalized Reduced Gradients, Bremicker (1986)
(8) CONLIN: Convex Linearization Method, Fleury (1989)
(9) QPRLT : Quadratic Programming with Reduced Linesearch Technique, Bremicker (1986)

Some theoretical aspects and experience show which strategy can be the best used for a certain problem. We will outline this point in the next chapter.
The modularity of $MBB - Lagrange$ makes it easy to add new strategies in the code.

6. Special Features

In this chapter we will call your attention to some special features of $MBB - Lagrange$ which make it a very interesting structural design system.
$MBB - Lagrange$ is suitable for large scale problems which occur in space and aircraft structures but also e.g. in automobile manufacturing. Especially in aircraft design problems the aeroelastic and flutter calculation is necessary. For the design of an aircraft the flutter speed is a decisive item. Flutter means an unstable vibration for a certain speed, the flutter speed. This speed must lie above the maximum flying speed of the aircraft. The flutter calculation is a very complex field in theoretical and numerical points of view.
Another highlight is the wide range of dynamics, see Ross (1988). In the code several eigenvalue solvers are implemented. Besides natural frequencies dynamic responses are to be taken into consideration: frequency, transient, and random response. A special feature is to restrict a range of frequencies. The user can choose if the structure response lies out of this range or within this range. If a certain eigenvalue is restricted $MBB - Lagrange$ gives the possibility to follow this mode (e.g. longitudinal vibration), even the order will change during the optimization process. This so called mode tracing is presented in the example (2). Another feature in this context is the possibility to define substructures for dynamics. Some elements build up a superelement for

which the eigen analysis will be performed. With this technic the calculation time can be reduced efficiently.

Concerning buckling it is possible to handle isotropic and composite material. Furthermore local stability of sandwich structures (wrinkling) can be taken into account which are very common in aerospace applications, cf. Dobler et al. (1991).

In some cases the so called system identification is useful. That means one can evaluate - approximately - the location of model imperfaction by taking measured data from modal tests.

Besides varying the thicknesses of layers it is possible to vary the layer angles as well (with fully analytical calculation of sensitivity). This facility leads to a further reduction of weight when defining both as design variables.

As mentioned above $MBB - Lagrange$ offers nine optimization algorithms. For the user it can be a problem to choose the right strategy for his special task. Indeed this is a difficult question. Therefore $MBB - Lagrange$ encloses an expert system (better called knowledge based system) which helps the user in his decision, see Schittkowski and Zotemantel (1990). After the input run the main data of the model are known. Then the user can start this knowledge based system and gets a list of all available strategies with so called safety factors. This factor indicates whether this strategy will be more or less successful: 0 means not possible, 100 means it is the best. The table shows an example for a typical middle scaled aircraft problem:

IBF	MOM	SLP	SRM	RQP1	RQP2	GRG	CONLIN
0	63	64	0	70	70	14	75

Table 1: Example for the choice of the strategy

The user defines a number of iterations to run the optimization process. If this optimization does not converge up to this limit or does not work at all the user can choose a restart or warmstart option. The program continues with the first choosen or another optimization algorithm beginning with the design last found.

Last but not least the code of $MBB - Lagrange$ is independent of computer operation systems. For this reason the code is easy to implement on different computer systems. The program is portable on VAX/VMS, VAX/ULTRIX, IBM/MVS, CRAY, and other UNIX systems, preferably it runs on workstations.

7. Application Examples

The following problems demonstrate the capacity of $MBB - Lagrange$ in various applications.

(1) Composite fin of a combat aircraft

This problem is typical for aircraft structures. The model of the whole aircraft and of the fin with the finite element mesh is shown in fig. 5.

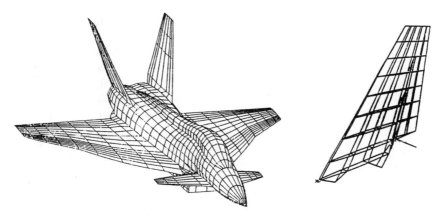

Figure 5: Finite element model of a combat aircraft and of its fin

In detail we study the composite fin with the stabilizer (4 layers) and the rudder (3 layers). This example will show the interdisciplinary simultaneous optimization to find the minimum weight. The design we will discuss here is restricted by stress and strain constraints, aeroelastic efficiencies and flutter speed. The optimization starts infeasible with uniform thickness distribution. Besides the thickness of the skin the layer angles can be design variables too.

A study demonstrates the variety of applications. The starting design is an optimized structure only with sizing variables (thicknesses, 102 design variables) which leads to a final weight of 42.3 kg (variable structure). The next step is to run a model with sizing plus 7 layer angle variables (linking for each layer of the stabilizer and rudder, resp.). Then the final weight is 34.6 kg. At least we arrive a final weight of 25.3 kg when every layer angle (in each element) of the first, the second, the third resp. the third and fourth layer is a design variable. Fig. 6 shows the thickness distribution and the orientation of the layers for this case.

(2) Vertical stabilizer of a helicopter

This concept study of a vertical stabilizer is interesting due to the two frequency constraints (besides stress constraints). The finite element model with the characteristic values is illustrated in fig. 7. The initial design is extremely violated (about 300 %), but the final design is feasible by reduction of weight (about 25 %), fig. 8.

One point must be mentioned here concerning the eigenmodes. In the initial design the first mode is a bending-twisting mode, the second is a pure bending mode. But in the final design it is the opposite: the first mode is now the pure bending mode and the second the bending-twisting mode. That means the modes have changed during the optimization process, fig. 9. This shows the importance of the mode tracing (see chapter 6).

(3) Horizontal stabilizer with endplates of a helicopter

Figure 6: Combat aircraft fin: optimal thickness distribution (sizing and layer angles)

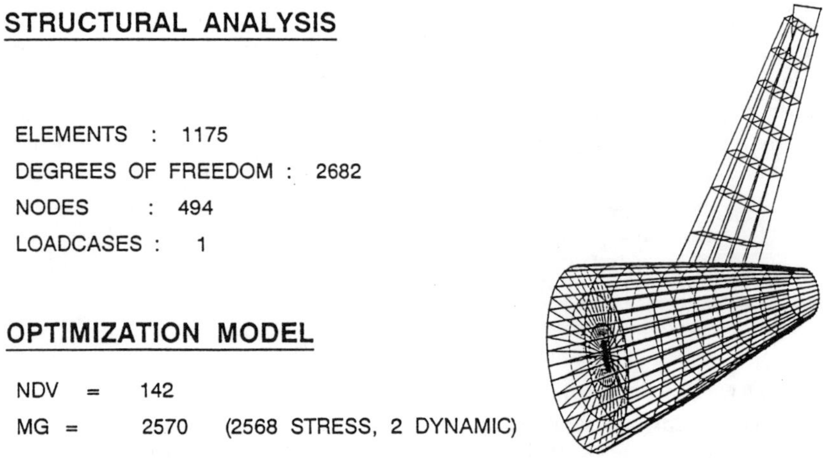

STRUCTURAL ANALYSIS

ELEMENTS : 1175
DEGREES OF FREEDOM : 2682
NODES : 494
LOADCASES : 1

OPTIMIZATION MODEL

NDV = 142
MG = 2570   (2568 STRESS, 2 DYNAMIC)

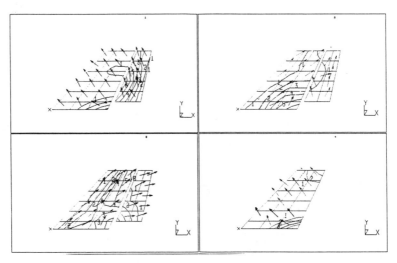

Figure 7: Finite element model of the vertical stabilizer of a helicopter

The structure consists of an airfoil section like an airplane wing and endplates which act as vertical stabilizers. The design of one half of the airfoil section is shown in fig. 10.

The upper and lower panels of the airfoil section are sandwich plates with a honeycomb core and aramid fiber face sheets. These sandwich plates and ribs transfer the aerodynamic loads to a spar. This spar is an I-shaped bar with straps. The endplates are sandwich plates of constant thickness.

Because of the symmetry of the structure only one half is idealized. Fig. 11 shows the fi-

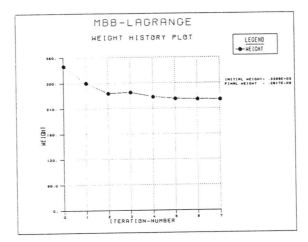

Figure 8: Weight history of the vertical stabilizer of a helicopter

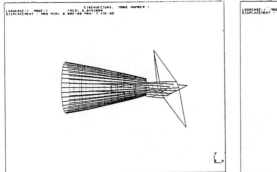 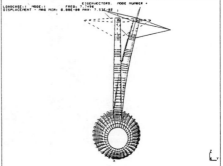

Figure 9: First mode of the initial and of the final design of the vertical stabilizer

nite element idealization with mainly composite membrane elements and the definition of design variables. To reduce the number of design variables and for manufacturing reasons the sizing variables are linked together to at least 144 variables. Three load cases define the loading of the stabilizer. The model has 597 DOFs and 3439 constraints. Besides constraints from sandwich wrinkling and composite failure criterion a lower bound for the first eigenfrequency is given.

The analysis should answer some questions. How is the interaction of weight reduction and the change of the first eigenfrequency? What is preferable: a thicker profile (15%) or the current design (12 %)? What yields the comparison of a structure with spar and a structure without spar?
To answer these questions the model has to modify from case to case. Different opti-

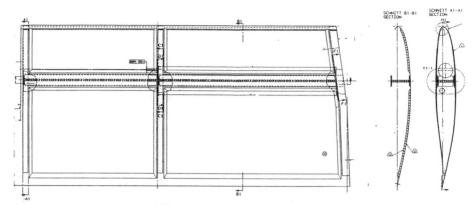

Figure 10: Horizontal stabilizer of a helicopter

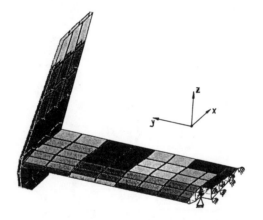

Figure 11: Finite element idealization of the horizontal stabilizer

mization runs are necessary. The results are shown in fig. 12.

The initial weight was 10.6 kg, and the first eigenfrequency was 14 Hz. Considering only strength restrictions a weight reduction of 3 kg is possible but the first eigenfrequency drops significantly. Holding the first eigenfrequency constant the weight decreases of 2 kg. Another result is that a design with spar (the current model) is far better than a design without it. Concerning the thickness of the airfoil section the result is that for low stiffness, i.e. a low first eigenfrequency, there is nearly no difference in the weight between a thin (12%) and a thick (15%) profile. However a thick profile is useful for high stiffness (high first eigenfrequency).

An optimization run on an IBM 3090 XA mainframe requires a total CPU time of 504 seconds.

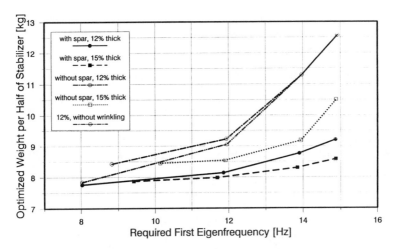

Figure 12: Minimum weight of the horizontal stabilizer versus the first eigenfrequency

8. Conclusion

This article has two aspects which the reader should mention: First it gives an overall view of the structural design system $MBB - Lagrange$. This software package is an optimization tool based on FEM and mathematical programming with special features with respect to space and aircraft structures. The highlights are the simultaneous optimization, a set of many different optimization algorithms, and an open program architecture. The application of large scale problems demonstrates how $MBB - Lagrange$ can help to solve complex problems.

The second aspect is to point out the importance of optimization for the designer. Optimization means the use of a computer aided tool supported by mathematical programming. To handle large and complex structures such tools are an aid for engineers to find optimal - or better - designs with reasonable costs. It is important to use all features of the CAE/CAD including structural optimization to get best possible results in engineering.

References

Eschenauer H. (1985), Rechnerische und experimentelle Untersuchungen zur Strukturoptimierung von Bauteilen. DFG-Forschungsbericht Universität Siegen IMR

Morris A.J. ed. (1982), Foundations of Structural Optimization: A Unified Approach. Wiley

Schuldt S.B. (1975), A method of multipliers for mathematical programming problems with equality and inequality constraints. Journal of Optimization Theory and Applications, Vol. 17 No. 1/2

Kneppe G. (1985), Direkte Lösungsstrategien zur Gestaltsoptimierung. VDI-Verlag

Haftka R.T., Kamat M.P. (1985), Elements of Structural Optimization. Martinus Nijhoff Publishers, The Hague

Schittkowski K. (1985/86), NLPQL: A FORTRAN subroutine solving constrained nonlinear programming problems, Annals of Operations Research, Vol. 5, 485-500

Powell M.J.D. (1984), The performance of two subroutines for constrained optimization on some difficult test problems. SIAM Conference of Numerical Optimization, DAMPT 1984/INAG

Bremicker M. (1986), Entwicklung eines Optimierungsalgorithmus der generalisierten reduzierten Gradienten, Anwendung auf Beispiele der Strukturdynamik Universität-GH Siegen, Institut für Mechanik und Regelungstechnik

Fleury C. (1989), CONLIN: An efficient dual optimizer based on convex approximation concepts. Structural Optimization, Vol. I, No. 2, Springer International

Ross C. (1988), Strukturoptimierung mit dynamischen Nebenbedingungen. Lehrstuhl B für Mechanik, Technical University of Munich

Dobler W., Erl P., Rapp H. (1991), Optimization of sandwich structures with respect to local instabilities with MBB-LAGRANGE. NATO/DFG Advanced Study Institute, Berchtesgaden

Schittkowski K., Zotemantel W. (1990), On a knowledge based user interface for the structural optimization system LAGRANGE, in: Artificial Intelligence in Computational Engineering, M. Kleiber ed., Ellis Horwood

Author's address:

Dr.-Ing. Rainer Zotemantel
Engineering Division
CAP debis IAS GmbH
Otto-Hahn-Str. 28
D-8012 Riemerling/München

The OASIS–ALADDIN Structural Optimization System

Björn Esping, Dan Holm and Odd Romell

Abstract. OASIS–ALADDIN is a comprehensive CAE package for optimization of mechanical structures in the linear response domain. The system allows the optimization of local and global geometry; membrane, shell, laminate and lamina thicknesses; cross–sectional area of rods; material orientation (e.g. fiber angels); and material property (e.g. sandwich foam core densities). Objective and constraint functions may be structural weight, mass moment of inertia, stiffness/deflections, natural frequency, stresses and material cost, as well as arbitrary linear combinations of these. Objective and constraint functions may also be defined as explicit functions of the design variables. The objective may be given as minimizing or maximizing a single function or a combination of several functions, or as minimizing the maximum of a set of functions. The shape optimization concept is based on CAD–formulations and is fully 3–D.

1. Introduction

1.1 General outline. OASIS–ALADDIN is a commercial code for optimization of mechanical structures in the linear response domain. It is a comprehensive package with the following important features: CAD-oriented modelling and post processing, parametric description of shapes, loads, etc., generalized design variables, capacity for many design variables (1000 or more), the Method of Moving Asymptotes (MMA), semianalytical derivatives, forward finite differences, exchangeable FE–code, IGES interface, menu driven user interface and compatibility with OCTOPUS. Typical applications include mechanical and structural design in engineering industries, e.g. automotive and aerospace.

1.2 Software organization. Structural optimization involves two distinct phases – formulation of the problem and the solution of it. OASIS–ALADDIN is organized in two parts: the ALADDIN pre/post processor and monitor part, and the OASIS optimization part. The two parts may reside on different hosts but OASIS is normally accessed and handled through ALADDIN.

OASIS is the iterative optimization part of the system. The analysis and modification loop is fully automatic and governed by the input data, but the user can monitor the process and has continuous access to intermediate results for inspection. The user can also intervene during the process to change a number of parameters, or to stop, modify and restart the process.

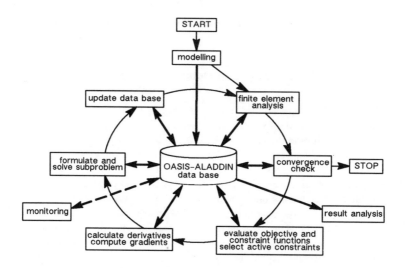

1.3 CAE integration, pre- and postprocessing. ALADDIN is the interactive pre and post processor, and provides the menu driven user interface of the system. Geometric modelling, meshing and generation of physical data (boundary conditions, loads, material properties a.s.o.), as well as the definition of design variables and formulation of the optimization problem, is done interactively in a CAD-type environment, so that the geometrical entities are central, rather than nodes and elements. All these entities have logical names, assigned to them by the user, which makes it easy to build well-organized models, even when they are large. These names are also useful when editing a model, as well as during result analysis. Thus, all input data are linked to the CAD-objects, and are therefore independent of the finite elements and nodes. For visualization and verification of allowed design variable intervals it is possible to animate the allowed shape variations. Pressures, thicknesses and prescribed displacements can be given as sculptured distributions over surfaces and lines, allowing complex distributions to be modelled. Geometric data can be imported from most CAD-systems as well as other FE pre-processors through the IGES format.

During the analysis the user has access to all intermediate results, for plotting and listing, as soon as they have been calculated. At any time during the process the user can intervene to change optimization parameters, design variable values or limits, type of approximation used in the optimization algorithms, or convergence criteria.

The post processing facilities allow presentation of stresses, strains, deflections and eigenfrequency modes, in isocolour or isoline plots with hidden lines removal; plots and tables of convergence histories of objective and constraint functions, design variable values, etc. The logical names of all geometric entities allow for selective and precise control over the result presentation.

1.4 Availability. User licenses for the OASIS–ALADDIN Structural Optimization System are available on an annual basis. The license fee includes installation media with instructions, documentation, basic support and introductory training at Alfgam's offices. Additional training and application courses are offered separately.

The system runs on a number of UNIX–based workstations, mainframe and super computer platforms as well as on VAX and VAXstation computers under VMS.

Engineering optimization services are also offered, on a case-to-case basis, as are dedicated developments for special applications.

1.5 Auxiliary. OASIS–ALADDIN may be integrated with the OCTOPUS system for concurrent, integrated multi-disciplinary optimization, where objective and constraint functions from different disciplines can be combined and treated simultaneously, e.g. shape optimization of ship hulls or car bodies including both structural and fluid mechanics responses, or production cost minimization with structural constraints. The system is especially designed for computer networks; communication between the different subjobs is handled through small ascii files, this makes the concept flexible and enables communication between all types of computers that can exchange ascii files. Evaluation of functions and gradients for each discipline can be executed on separate computers, the values are transferred to a central optimizer once per iteration. The communication may be administered from a separate 'master' computer which can be rather small (workstation) even if the analyses are of super computer size. The system has an open architecture so that user written programs, as well as third party systems, can be integrated. It can handle up to 100 different types of optimization jobs simultaneously. OCTOPUS is now in commercial use, and has proven its usefulness in a number of applications.

Compared to an ordinary "purely structural" optimization the scope is significantly widened, and several obvious advantages are gained, e.g.:

- a wider selection of properties can be included in the optimization
- the user may include specialized in-house code in the optimization
- different sub-analyses may be run independently (in parallel and/or sequentially) on the type of hardware and at the location that is best suited
- separate sub-analyses of the same type can be run concurrently rather than sequentially, and each with the appropriate accuracy and discretization.

2. Optimization Model

It is of great importance that the optimization problem is correctly formulated, since the optimizer uses every possibility allowed to improve the objective, even possibilities that were not intended by the user. In the problem formulation the following main decisions have to be made:

1: what is the *objective function*?
2: what *constraints* are necessary?
3: and what type(s) of *design variables* should be used.

The objective and the constraint functions are functions of a set of *design variables* x_j. There are: sizing variables, shape variables, material orientation variables and material property variables. The sizing variables are often *continuous* but for certain problems they have to take *discrete* values. This may be the case for e.g. composite material structures where the design variables may be the number of plies in a certain lay up direction.

2.1 Mathematical formulation. The objective function can be formulated either as minimizing a linear combination of a number of response functions or as minimizing the maximum of a group of candidate functions. The structure is subject to a number of constraints, e.g. displacements, stresses, etc. The constraints together with the objective function form an optimization problem P. In mathematical terms the problem is often stated as:

$$
\begin{aligned}
\text{P:} \quad & \text{minimize:} \quad w(x) & & \text{objective function} \\
& \text{subject to:} \quad g_i(x) \leq \overline{g}_i & , i=1,I & \quad \text{constraints} \\
& \qquad \underline{x}_j \leq x_j \leq \overline{x}_j & , j=1,J1 & \quad \text{continuous design variables} \\
& \qquad (x_{1j}, x_{2j}, x_{3j}, \dots) & , j=J1+1,J & \quad \text{discrete variables}
\end{aligned}
$$

where $w(x)$ is the objective function, and x_j, $j=1,J$ is the set of design variables. $\underline{x}_j$ and $\overline{x}_j$ are the lower and upper limits, respectively, for the continuous design variables and $(x_{1j}, x_{2j}, x_{3j}, \dots)$ are the allowed values for discrete variables. $g_i(x)$, $i=1,I$ are the constraints and $\overline{g}_i$ their limits. I is the number of constraints and J is the total number of design variables. P is typically a non-linear optimization problem since in general both the objective function w and the constraints g_i are implicit, non-linear functions of the design variables x_j.

OASIS normally solves the following problem:

$$P: \quad \min \quad W = \Sigma C_i \cdot g_i \quad ; i = 1, I1$$

$$\text{subject to:} \quad g_i \leq 1 \quad ; i = I1+1, I$$

$$\underline{x}_j \leq x_j \leq \overline{x}_j \quad ; j = 1, J$$

The functions g_i, $i = 1, I1$ are constraints defined by user input. They are used internally on their standard form $g_i \leq 1$, i.e. for a deformation $g_i = d/\overline{d}$ etc. The composition of W and the values for C_i are defined by user input.

OASIS can also solve the min-max problem:

$$P: \quad \min \quad (\max (C_i \cdot g_i)) \quad ; i = 1, I1$$

$$\text{subject to:} \quad g_i \leq 1 \quad ; i = I1+1, I$$

$$\underline{x}_j \leq x_j \leq \overline{x}_j \quad ; j = 1, J$$

Any g_i can be included in the combined objective function W or the min-max problem. A typical example of min-max problems is the minimization of a stress concentration in a certain domain of the structure, or the maximization of the lowest eigenfrequency. The min-max option can be used with MMA for problems with continuous design variables.

The optimization problem is solved iteratively. In each iteration the original problem is approximated by a subproblem which is solved. The approximation requires information on the values of the objective and constraint functions as well as their derivatives with respect to the design variables. This information is calculated using finite element techniques, semi-analytical derivatives and finite forward differences. Starting from a design proposal (or an existing design) the automatic and iterative process of analysis and modification will improve the design until a convergent solution, or a solution that satisfies the designer, is found. The iteration procedure is discontinued when either the number of iterations reaches its given upper limit, or the solution has converged. The maximum number of iterations and the tolerances for objective function convergence and solution feasibility are optional inputs.

P may often be non-convex and have multiple minima. A *constraint surface* ($g_i = \bar{g}_i$) divides the design space into a *feasible* and an *infeasible* region. If the line, connecting two points on the constraint surface, is in the feasible region then the constraint is *convex*. The *gradient* of the constraint ∇g_i is the direction in which we have the greatest increase of g_j. ∇g_i is 'normal' to the constraint surface. If the constraint surface is not convex we get several minima. The *global minimum* or *optimum* can be hard to distinguish. Current solution methods will only find a local minimum which is not necessarily the global optimum. Our experience is that problems with sizing variables will give rather convex problems, while problems with material orientation variables will often give non-convex problems. Shape variable problems can be of both sorts. In a problem with multiple minima the location of the starting point will often determine which of the local minima will found.

2.2 The parametric formulation. The structure to be optimized is defined by a set of basic bodies such as lines, surfaces and solids. All bodies originate from a set of points which can be connected in various ways. Basic bodies are combined to form a complex structures. The shape description within each body is parametric which means that a point in the parameter space (ξ,η,ζ) corresponds to a point $r = (x,y,z)$ in the Euclidean space (fig. 2.1). All parameters have values in the range [0,1].

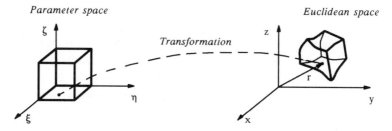

Fig. 2.1 *Transformation from parameter space to Euclidean space*

The transformation is done by a mapping function F. We express the relation as:

$$r = F(\xi,\eta,\zeta), \quad \xi \in (0,1), \quad \eta \in (0,1), \quad \zeta \in (0,1)$$

Let us as an example take a parabolic line: The parabolic line connects three points p_0, p_1 and p_2, which have the Euclidean coordinate vectors r_0, r_1 and r_2. For a line, the parameter space is always one dimensional (ξ). The parameter values for p_0, p_1 and p_2 are 0, 0.5 and 1, respectively. The mapping function F is as follows:

$$r = 2(1-\xi)(0.5-\xi)r_0 + 4\xi(1-\xi)r_1 + 2\xi(\xi-0.5)r_2$$

Notice that for a fix parameter value ξ, r is a function of r_0, r_1 and r_2 only. We get:

$$r = F(\xi,r_j) \qquad ; j = 0,1,2$$

Each body has its own mapping function F_i and there is no mapping interaction between different bodies. Notice that if a point p_j is moved, all interior points (nodes) will follow smoothly. What we do now is to attach design variables to any of the coordinates r_j in a point p_j. It follows that very few design variables can have great impact on the shape.

2.3 Mesh generation. The analysis requires a computation (finite element) mesh. This can easily be generated in the parameter space. The parametric mesh is then mapped, using the transformation function F, to the Euclidean space (fig. 2.2).

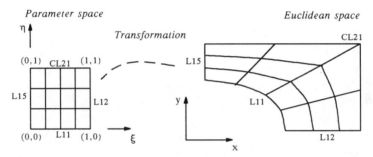

Fig. 2.2 Mapping

The shape optimization concept is based on general parametric 3-D CAD type formulations. Shape variables x are attached to the points. A change of a design variable will influence the entire mesh belonging to the body of which the point is a part. Consider for a moment a structure defined for example by a Coons surface. If we change the value of one of the design variables x (i.e. a point coordinate), by a small step Δx we get for the same set of parameter values (ξ,η), a new mesh which is a smooth variation of the original mesh. Later we will see how this gives us the necessary derivatives of the response functions with respect to the design variables.

2.4 Objective and constraint functions. Objective and constraint functions are initially formulated in the same way, the objective function is then selected from this group. Several functions may be combined to form the objective function. The following quantities can be handled as objective and constraint functions:

- structural weight $\qquad (w \leq \bar{w})$
- material cost $\qquad (c \leq \bar{c})$
- moment of inertia $\qquad (I \leq \bar{I})$
- displacement $\qquad (d \leq \bar{d})$
- eigenfrequency $\qquad (\omega \geq \underline{\omega})$

- stresses $\quad(T_s \leq 1)$
- explicit linear combinations of design variables $\quad(e_L \leq \bar{e}_L)$

The evaluation of the functions and their derivatives will be described later.

2.5 Design variables. The design variables in OASIS–ALADDIN allow the optimization of global and local geometry; membrane, shell, laminate and lamina thicknesses; material orientation (e.g. fiber angels); material selection (e.g. sandwich foam core densities) and cross-sectional area of rods. Design variables can be continuous or discrete, and combinations of both types can be used simultaneously. Discrete variables are given either as an allowed interval with a step size, or as a table of explicit values. The material property variables are discrete variables that allow the system to select one of several materials from a table supplied by the user. A material orientation variable is explicitly the angle between the principal material direction and a local or global reference direction vector. Shape variables may be combined to define slave variables, and may also be used as passive design parameters.

2.6 Variable linking. All design variables, as well as loads, boundary conditions, thicknesses, materials, etc., are normally linked to the geometry model. This makes the optimization model independent of the FE mesh and the model may be edited without affecting the mesh, or the discretization changed without need to alter the geometry model. Of course it is still possible to apply loads, etc., to the FE model instead, and to use a combination of both methods.

3. Structural Analysis

The OASIS–ALADDIN Structural Optimization System as standard contains a general purpose Finite Element package for linear static analysis of structural properties and responses such as stress, strain, displacement, eigenfrequency and so on. The FE system is also used to calculate the structural matrices for design variable perturbations $(x+\Delta x)$ for the sensitivity analysis. For special applications other FEM packages may be substituted.

3.1 Finite elements. The finite element library contains linear and quadratic analytical rod elements, and quadratic isoparametric shells, membranes and solids. Membranes and shells may be stacked and there is a special sandwich element. Other special elements include contact elements and surface stress elements.

The 8-node quadrilateral membrane element is a plane, constant thickness isoparametric element, with two degrees of freedom per node. It may be collapsed into a 6-node triangle. Numerical integration is performed with the Gauss integration method.

The 8-node shell element is a doubly curved, variable thickness isoparametric element, generally referred to as an Ahmad element. Transverse shear is taken into account and each node has five degrees of freedom. The element is transformed into a normalized plane with coordinates ξ and η in the interval $[-1,1]$. Numerical integration is performed in the η-ξ plane with the Gauss integration method. For homogeneous shell elements, two points are used through the thickness. For stacked shell elements, an equivalent stiffness is calculated. Stresses are given on both surfaces at the Gauss points. For a stacked element, the stresses are computed on both sides of each layer at the Gauss points.

Both membrane and shell elements can be stacked, i.e. composed of an arbitrary number of orthotropic plies, where each ply has an individual thickness t_k. Independent thickness and material property variables can be connected to each layer. A different material direction can be defined for each ply. Material orientation variables will apply to the entire stack, rotating it while keeping the relative orientation of the individual plies constant.

The sandwich element is an 8-node doubly curved isoparametric element. The shell element automatically becomes a sandwich element if it is stacked and if *one* of the central plies has shear moduli G23, G31 which are less than 10% of the corresponding values of surrounding plies.

The volume element is a 20-node quadratic isoparametric brick element. It may be collapsed into a 15-node wedge element.

The material models include orthotropic plane strain and plane stress materials for membranes, as well as orthotropic volume materials for shells and solids. The principal material directions are defined by reference direction vectors given by the user, which may be linked to an orientation design variable, or by the local coordinate system of the element.

3.2 Loads and boundary conditions. Loads can be entered as point loads, line loads, pressures, gravity and centrifugal loads, and as prescribed displacements. Pressures are generalized to include force/length, force/area and force/volume. Distributed loads (pressures) and boundary conditions can be sculptured by using the shape functions to interpolate load values given at the control points, i.e. a line load applied to a parabolic line will have a parabolic distribution a.s.o. Internally all loads are recalculated and applied as consistent node point forces.

The deflection in any degree of freedom and for any node can be coupled to the deflection of any other node. The deflection of the dependent node is defined by a linear combination of the deflections of an arbitrary number of independent nodes.

3.3 Failure Criterion. Output includes principal stress and strain values. The element stresses are normally used to calculate an effective stress (von Mises), or to calculate a normalized failure factor. The failure factor Ts is calculated according to the Tsai-Hill formula. For isotropic materials the Tsai-Hill failure factor is equivalent to the von Mises stress.

4. Sensitivity Analysis

4.1 General strategy. The gradient calculations are based on semianalytical derivatives and forward finite differences (ffd), which gives a high computational efficiency. With this approach the computational cost increase is only linear in the number of design variables, which allows a very large number of variables, 1000 or more, without prohibitive computational cost.

4.2 Constraint selection. To make the evaluation of the gradients as efficient as possible only the most critical constraints in each iteration are active. The selection is done slightly different for the different types of constraints. For displacement constraints we select for the first iteration (k=1) a number of the most critical constraints, i.e. those with the largest values of g_i. For k > 1 we select those constraints which were active in the preceding iteration, i.e. those for which $\lambda_i^{(k-1)} > 1$, where λ_i is the Langrangian multiplier corresponding to constraint g_i. In addition to these, we select among the remaining a certain number of the most critical ones. For strain-stress constraints we do similarly in selecting active constraints for each iteration. The remaining strain-stress constraints can be used to update the lower bounds of the thickness and cross section area variables. For eigenfrequency constraints OASIS selects a given number of the lowest frequencies in each iteration to check that none of them is below the limit.. This number and the number of selected displacement and stress constraints as well as the criteria for what is to be considered as critical constraints may be input by the user.

4.3 Structural weight and material cost. The total weight of a structure is the sum of the weights of all the finite elements:

$$w = \sum_e w_e \qquad (4.3.1)$$

The constraint written on the standard

form is:
$$g = \frac{w}{\bar{w}} \leq 1 \quad (\bar{w} > 0) \tag{4.3.2}$$

Its derivative is:
$$\frac{\delta g}{\delta x} = \frac{\delta w}{\delta x} / \bar{w} \tag{4.3.3}$$

where
$$\frac{\delta w}{\delta x} = \sum_e \frac{\delta w_e}{\delta x} \tag{4.3.4}$$

Now, $\delta w_e/\delta x \neq 0$ only if x is a design variable active in that element. The derivative is calculated numerically using ffd:

$$\frac{\delta w_e}{\delta x} = \frac{w_e(x + \Delta x) - w_e(x)}{\Delta x} \tag{4.3.5}$$

where $w_e(x)$ and $w_e(x+\Delta x)$ are calculated by the FE-system. Derivatives of all other element matrices are treated similarly. Of course the step Δx has to be small.

The material cost of the structure may be though of as a kind of density ($/kg) and is treated in the same way as the structural weight.

4.4 Moment of inertia. We write the total moment of inertia as:

$$I = \sum_e I_e \qquad \text{where} \qquad I_e = \sum_n m_n \cdot r_n^2$$

and the summation is done over all node points n within element e and m_n are the corresponding masses. The standard form for the constraint g is:

$$g = \frac{\bar{I}}{I} \leq 1 \quad (\bar{I} > 0)$$

The derivatives are calculated according to:
$$\frac{\delta g}{\delta x} = \frac{\delta I}{\delta x} / \bar{I}$$

where
$$\frac{\delta I}{\delta x} = \sum_e \frac{\delta I_e}{\delta x}$$

ffd:
$$\frac{\delta I_e}{\delta x} = \frac{I_e(x + \Delta x) - I_e(x)}{\Delta x}$$

The masses m_n are computed by the FE-system for variable values x and $x+\Delta x$.

4.5 Displacements. The FE-problem can be formulated:

$$Ku = p \tag{4.5.1}$$

where K is the system stiffness matrix, u is the displacement vector and p the load vector. We can express a specific displacement d as a linear combination of u:

$$d = u^T q \tag{4.5.2}$$

where q is a vector consisting of coefficients c_j at locations corresponding to selected degrees of freedom in the structure. q will later be considered as a virtual load in the FE-problem.

The normalized displacement constraint g is:

$$g = \begin{cases} \dfrac{d}{\bar{d}} \leq 1 & (\bar{d} > 0) \\ d + 1 \leq 1 & (\bar{d} = 0) \end{cases} \qquad (4.5.3)$$

Its derivative is:

$$\frac{\delta g}{\delta x} = \begin{cases} \dfrac{\delta d}{\delta x}\Big/\bar{d} & (\bar{d} > 0) \\ \dfrac{\delta d}{\delta x} & (\bar{d} = 0) \end{cases} \qquad (4.5.4)$$

If we differentiate (4.5.2) and use (4.5.1) we get:

$$\frac{\delta d}{\delta x} = \left(\frac{\delta p^T}{\delta x} - u^T \frac{\delta K}{\delta x}\right) K^{-1} q + u^T \frac{\delta q}{\delta x} \qquad (4.5.5)$$

Introduce the virtual displacement v:

$$Kv = q \implies v = K^{-1} q \qquad (4.5.6)$$

Use (4.5.6) and notice that $\delta q/\delta x = 0$.

$$\frac{\delta d}{\delta x} = \frac{\delta p^T}{\delta x} v - u^T \frac{\delta K}{\delta x} v \qquad (4.5.7)$$

K and p are assembled from the contributions of all the elements, i.e. k_e, p_e, etc.

$$\frac{\delta d}{\delta x} = \sum_e \frac{\delta p_e^T}{\delta x} v_e - \sum_e u_e^T \frac{\delta k_e}{\delta x} v_e \qquad (4.5.8)$$

Again, $\delta p_e/\delta x$ and $\delta k_e/\delta x \neq 0$ only if x is a design variable within that element, and again we use ffd:

for element loads:

$$\frac{\delta p_e}{\delta x} = \frac{p_e(x + \Delta x) - p_e(x)}{\Delta x} \qquad (4.5.9)$$

for element stiffnesses:

$$\frac{\delta k_e}{\delta x} = \frac{k_e(x + \Delta x) - k_e(x)}{\Delta x} \qquad (4.5.10)$$

where $k_e(x)$, $k_e(x+\Delta x)$, $p_e(x)$, $p_e(x+\Delta x)$ are calculated by the FE-system. Equation (4.5.6) is also solved by the FE-system.

4.6 Eigenfrequency. The eigenvalue problem can be formulated:

The OASIS-ALADDIN structural optimization system

$$(K - \omega^2 M)a = 0 \quad (4.6.1)$$

where M is the structural mass matrix and a is the eigenvector corresponding to the eigenvalue ω.

We can state the constraint as: $\quad \omega^2 \geq \underline{\omega}^2 \quad (\underline{\omega} > 0) \quad (4.6.2)$

which when written on its standard form becomes:

$$g = 2 - (\frac{\underline{\omega}}{\omega})^2 \leq 1 \quad (4.6.3)$$

Its derivative is: $\quad \dfrac{\delta g}{\delta x} = -\dfrac{\delta \omega^2}{\delta x} / \omega^2 \quad (4.6.4)$

Differentiate (4.6.1): $\quad (\dfrac{\delta K}{\delta x} - \dfrac{\delta \omega^2}{\delta x} M - \omega^2 \dfrac{\delta M}{\delta x})a + (K - \omega^2 M)\dfrac{\delta a}{\delta x} = 0 \quad (4.6.5)$

and premultiply with a^T:

$$a^T \dfrac{\delta K}{\delta x} a - \dfrac{\delta \omega^2}{\delta x} a^T M a - \omega^2 a^T \dfrac{\delta M}{\delta x} a + a^T (K - \omega^2 M) \dfrac{\delta a}{\delta x} = 0 \quad (4.6.6)$$

Assume that a is normalized with respect to M, i.e. $a^T M a = 1$, use (4.6.1) and transform (4.6.6) to the element level. We then get:

$$\dfrac{\delta \omega^2}{\delta x} = \sum_e a_e^T \dfrac{\delta k_e}{\delta x} a - \omega^2 \sum_e a_e^T \dfrac{\delta m_e}{\delta x} a \quad (4.6.7)$$

$\dfrac{\delta k_e}{\delta x}$ is given by (4.5.10) and derivative of the the element mass matrix is found, again by ffd as: $\quad \dfrac{\delta m_e}{\delta x} = \dfrac{m_e(x + \Delta x) - m_e(x)}{\Delta x} \quad (4.6.8)$

where $m_e(x)$, and $m_e(x+\Delta x)$ are calculated by the FE-system.

4.7 Strain, stress or equivalent stress. Stresses are always calculated at the integration points. If the stresses are required on the surface of a solid or the edge of a 2-D element, other elements calculating these stresses can be included. Before we calculate the equivalent stress we must have expressions for the strain and stress vectors. The strain vector ϵ of an element expressed in its element local coordinate system is:

$$\epsilon^T = u^T q \quad (4.7.1)$$

where q is the strain-displacement matrix for the finite element. Depending on the element type we are considering we get:

rod: $\quad \epsilon = \epsilon_x \quad (4.7.2)$

membrane: $\quad \epsilon = (\epsilon_x, \epsilon_y, \gamma_{xy})^T$ (4.7.3)
shell and sandwich: $\quad \epsilon = (\epsilon_A, \epsilon_B)^T$ (4.7.4)
solid: $\quad \epsilon = (\epsilon_x, \epsilon_y, \epsilon_z, \gamma_{xy}, \gamma_{yz}, \gamma_{xz})^T$ (4.7.5)

ϵ_x, ϵ_y, ϵ_z, γ_{xy}, etc are the strain components expressed in the local x, y and z axes. Let us explain ϵ_A, and ϵ_B. Let ζ be a non-dimensional z coordinate.

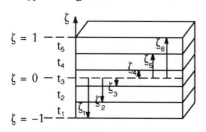

Fig. 4.1 Multilayer shell

The strain as a function of ζ can be expressed as: $\quad \epsilon_\zeta = \epsilon_A + \zeta \cdot \epsilon_B$ (4.7.6)
where ζ for the lower and upper side of each ply (see Fig 4.1) are:

$$\zeta_i = 2 \cdot \frac{\sum_{j=1}^{i-1} t_j}{\sum_{j=1}^{n} t_j} - 1$$

(4.7.7)

and n is the number of plies.

$$\epsilon_A = (\epsilon_x, \epsilon_y, \epsilon_z, \gamma_{xy}, \gamma_{yz}, \gamma_{xz})_{\zeta=0}$$ (4.7.8)

and ϵ_B is the corresponding "bending" contribution.

Differentiate (4.7.1), use (4.5.1) and proceed as for the displacement constraints:

$$\frac{\delta \epsilon^T}{\delta x} = \sum_e \frac{\delta p_e^T}{\delta x} v_e - \sum_e u_e^T \frac{\delta k_e}{\delta x} v_e + u^T \frac{\delta q}{\delta x}$$ (4.7.9)

where the virtual displacement v is: $\quad Kv = q$ (4.7.10)

(v_e is its subset for element e) and q is given by (4.7.1). The derivatives $\delta p_e/\delta x$ and $\delta k_e/\delta x$ are given by (4.5.9) and (4.5.10) while

$$\frac{\delta q}{\delta x} = \frac{q(x + \Delta x) - q(x)}{\Delta x}$$ (4.7.11)

where q(x) and q(x + Δx) are calculated by the FE-system. In the shell or sandwich case we have to express the derivative of ϵ_ζ as:

$$\frac{\delta \epsilon_\zeta}{\delta x} = \frac{\delta \epsilon_A}{\delta x} + \frac{\delta \zeta}{\delta x} \epsilon_B + \zeta \frac{\delta \epsilon_B}{\delta x} \qquad (4.7.12)$$

where $\delta \epsilon_A/\delta x$ and $\delta \epsilon_B/\delta x$ are subvectors of $\delta \epsilon/\delta x$ (see 4.7.4). It remains to find $\delta \zeta/\delta x$. But $\delta \zeta/\delta x \neq 0$ only if $x = t_k$ (a ply thickness), when we get:

$$\frac{\delta \zeta}{\delta t_k} = \begin{cases} 2 \cdot \dfrac{\sum_{j=1}^{n} t_j - \sum_{j=1}^{i-1} t_j}{(\sum_{j=1}^{n} t_j)^2} & i-1 \geq k \\[2ex] -2 \cdot \dfrac{\sum_{j=1}^{i-1} t_j}{(\sum_{j=1}^{n} t_j)^2} & i-1 < k \end{cases} \qquad (4.7.13)$$

In the remaining part of this chapter, we will drop the index of ϵ_ζ. From now on ϵ will in the sandwich or shell case mean ϵ_ζ as defined by (4.7.6). The strain vector ϵ_k, in a certain ply k (lower or upper side), expressed in its main material directions, can be written as:

$$\epsilon_k = T_k \cdot \epsilon \qquad (4.7.14)$$

$T_k = T(\theta_k)$ is a transformation matrix and $\theta_k = \phi + \beta_k$. In the rod case T_k is a scalar, i.e. $T_k = 1$. T_k is usually defined in the FE-system but can as an alternative be calculated by OASIS.

For the membrane case we have:

$$T_k = \begin{bmatrix} \cos^2 \theta_k & \sin^2 \theta_k & \sin \theta_k \cos \theta_k \\ \sin^2 \theta_k & \cos^2 \theta_k & -\sin \theta_k \cos \theta_k \\ -2 \sin \theta_k \cos \theta_k & 2 \sin \theta_k \cos \theta_k & \cos^2 \theta_k - \sin^2 \theta_k \end{bmatrix} \qquad (4.7.15)$$

Differentiate (4.7.14): $\qquad \dfrac{\delta \epsilon_k}{\delta x} = \dfrac{\delta T_k}{\delta x} \epsilon + T_k \dfrac{\delta \epsilon}{\delta x} \qquad (4.7.16)$

where $\delta \epsilon/\delta x$ is given in (4.7.9) and $\delta T_k/\delta x$ is calculated using FFD. The stress vector σ_k corresponding to ϵ_k (of ply k) is calculated as:

$$\sigma_k = C_k (\epsilon_k - \theta \cdot \alpha_k) \qquad (4.7.17)$$

where C_k is the appropriate constitutive matrix (Hooke matrix), α_k the vector of thermal expansion coefficients, θ is the temperature and ϵ_k is given by (4.7.14).

Differentiate (4.7.17):

$$\frac{\delta \sigma_k}{\delta x} = \frac{\delta C_k}{\delta x}(\epsilon_k - \theta \cdot \alpha_k) + C_k \cdot (\frac{\delta \epsilon_k}{\delta x} - \frac{\delta \theta}{\delta x}\alpha_k - \theta \frac{\delta \alpha_k}{\delta x}) \qquad (4.7.18)$$

Note that $\delta C_k/\delta x$ and $\delta \alpha_k/\delta x$ are non-zero only for material property variables. They are calculated using finite differences (fd) directly in the curves $C_k(mno)$ and $\alpha_k(mno)$. In those curves, derived as splines through the values in the material tables supplied by the user, the (discrete) design variable value represents the material number (mno).

Usually we constrain an equivalent stress. The complete derivation is too lengthy to present here, suffice it to take a quick glance at the normalized equivalent stress (Tsai factor) for the 2-D membrane case:

$$Ts_k = ((\frac{\sigma_1}{S1})^2 - \frac{1}{RHO}(\frac{\sigma_1 \cdot \sigma_2}{S1 \cdot S2}) + (\frac{\sigma_2}{S2})^2 + (\frac{\tau_{12}}{S12})^2)_k^{1/2} \qquad (4.7.19)$$

where $\sigma_k = (\sigma_1, \sigma_2, \tau_{12})_k^T$ and σ_1, σ_2 and τ_{12} are the stresses in the main material directions and S1, S2, S12 are the strengths in unidirectional tension or compression in corresponding directions. The interaction factor RHO is usually defined as S1/S2. The Tsai factor is similarly defined for shells and solids. The derivatives of the strength properties S1, S2, S12 and RHO are calculated in a similar manner as those of C and α.

In principle constraints can be put on any of the components in ϵ_k, σ_k or Ts_k. The constraint on Ts_k will finally become:

$$g = Ts_k \leq 1 \qquad (4.7.20)$$

For the complete derivation see Esping (1986a)

5. Optimization algorithms

The optimization problem can be solved in mainly two ways:
    1: optimality criterion methods or
    2: mathematical programming methods.

5.1 Optimality criterion methods. The optimality criterion (OC) methods postulate a criterion for optimum. This criterion is then used to create recursive formulas. The most commonly used criterion is the Fully Stressed Design (FSD) criterion. The FSD criterion is based on the assumption that at optimum every structural member is critically stressed in at least one loading case. FSD is obviously only useful for stress constraints and is only used for sizing problems. OASIS–ALADDIN can use the FSD criterion to update cross section areas and thicknesses. The recursive formula used for the cross sectional area x_j in a rod is called the *stress ratio method*:

$$x_j^{(k+1)} = x_j^{(k)} \cdot \frac{\sigma_j^{(k)}}{\bar{\sigma}_j}$$

where $\sigma_j^{(k)}$ is the most critical stress over all loading cases in the same element. $\bar{\sigma}_j$ is the maximum allowable stress (the strength) and k the iteration number. The FSD corresponds to a vertex in the design space which does not always correspond to the optimum design. The stress ratio method is however inexpensive and usually gives very satisfactory results.

5.2 Mathematical programming methods are based on the *Kuhn–Tucker* (KT) criterion which states that, at optimum, the following conditions are satisfied:

$$\nabla w + \sum_i \lambda_i \nabla g_i = 0$$

$$\lambda_i (g_i - \bar{g}_i) = 0 \qquad i = 1, I$$

$$\lambda_i \geq 0 \qquad j = 1, J$$

∇w and ∇g_i are the gradients of w and g_i in a design point x. We can for instance express ∇w as:

$$\nabla w = \left(\frac{\delta w}{\delta x_1}, \frac{\delta w}{\delta x_2}, \frac{\delta w}{\delta x_3}, \ldots \right)^T$$

The KT criterion says that an optimum is reached when the gradient of the objective function is a linear combination of the active constraint gradients. Active are those constraints for which $g_i = \bar{g}_i$ and $\lambda_i > 0$, where λ_i are the Lagrangian multipliers.

5.3 Approximation schemes. The mathematical programming methods require many function and gradient evaluations. Those evaluations are quite expensive as each evaluation involves an FE–analysis. An alternative approach is to create a sequence of explicit approximative *subproblems* $\widetilde{P}^{(k)}$ which converges to the solution of P. Each subproblem requires only one evaluation of the original objective and constraint functions and their derivatives. $\widetilde{P}^{(k)}$ is of a rather simple form and can be solved very efficiently. The number of necessary subproblems to achieve convergence is in general rather small but is strongly dependent on the quality of the chosen approximations. The cost to create $\widetilde{P}^{(k)}$ is high but the cost to solve it is low. OASIS_ALADDIN allows the user to choose between a combined linear/inverse approximation and the method of moving asymptotes.

In the *linear approximation*, LA, both the objective function w and the constraints g_i are approximated by a first order Taylor expansion around the preceding design point $x^{(k)}$:

$$\widetilde{P}^{(k)}: \min \widetilde{w}(x) = w^{(k)} + \sum_j \left(\frac{\delta w}{\delta x_j}\right)^{(k)} (x_j - x_j^{(k)})$$

subject to:
$$\widetilde{g}_i(x) = g_i^{(k)} + \sum_j \left(\frac{\delta g_i}{\delta x_j}\right)^{(k)} (x_j - x_j^{(k)}) \leq \bar{g}_i \; ; \; i = 1, I$$

$$\underline{x}_j \leq x_j \leq \bar{x}_j \qquad ; \; j = 1, J$$

This is a regular linear programming (LP) problem which can be solved by the Simplex method. Let us look at a subproblem $\widetilde{P}^{(k)}$. We assume that $w(x)$ is linear. The approximate subproblem $\widetilde{P}^{(k)}$ is marked by dashed lines and its solution is the design point $x^{(k+1)}$.

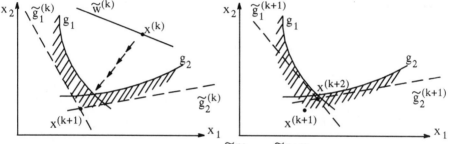

Fig. 5.1 Linear approximation of subproblems $\widetilde{P}^{(k)}$ and $\widetilde{P}^{(k+1)}$.

The procedure is repeated around $x^{(k+1)}$. The new solution is $x^{(k+2)}$. We see that the solutions $x^{(k)}$, $x^{(k+1)}$ and $x^{(k+2)}$ are moving closer to the optimum of P.

The sequence of linear approximations (LA) may result in an unstable convergence. This is often the case when we have very few active constraints. The convergence can be improved if we introduce move limits or *boxes* around preceding design points.

We have: $\quad x_j^{(k)} - h_j^{(k)} \leq x_j \leq x_j^{(k)} + h_j^{(k)}$

where $\quad h_j^{(k)} = \dfrac{c_j}{k + d_j}$

c_j and d_j define the box and may be given as optional input by the user. Default is: $c_j = \bar{x}_j - \underline{x}_j$ and $d_j = 1.0$. For thickness variables we always have:

$$\frac{1}{2} x_j^{(k)} \leq x_j \leq 2 x_j^{(k)}$$

The OASIS-ALADDIN structural optimization system

In all cases the given side constraints ($\underline{x}_j \leq x_j \leq \bar{x}_j$) are also valid.

Now, the FSD method can be considered to be a zero-order approximation. These approximations correspond to the side constraints and can be combined with first order approximations of other constraints. Thus it is possible to combine the two methods (FSD and LA). The convergence, however, is slow.

For sizing problems, the quality of $\widetilde{P}^{(k)}$ can be improved if the constraints, at least stress and displacement constraints, are linearized in the reciprocal variables $1/x_j$ instead of in the variables x_j themselves. This is the *inverse approximation*, IA. The inverse approximation is well suited for pure sizing problems but lacks generality. The linear approximation is general but its convergence properties are unsatisfactory. A conservative convex hybrid of the two approximations, HA, has been proposed by Haftka:

$$\text{linear in } x_j \text{ if } \frac{\delta g_i}{\delta x_j} > 0$$
$$\text{linear in } 1/x_j \text{ if } \frac{\delta g_i}{\delta x_j} \leq 0$$

Any given objective or constraint function may have a linear approximation with respect to one design variable and inverse with respect to another. When LA/IA is activated all variable types except sizing variables will be approximated according to the linear approximation, LA, while sizing variables (thickness and cross sections) are approximated using IA.

5.4 Method of moving asymptotes. A further generalization of HA has been proposed by Svanberg (1987, 1991). It is called the *Method of Moving Asymptotes* (MMA). In the method of moving asymptotes we introduce the new variables:

$$y_j = 1/(x_j - L_j) \text{ and}$$
$$z_j = 1/(U_j - x_j)$$

We notice that $L_j \to 0 \Rightarrow y_j \to 1/x_j$ and it can be shown that for $x_j \ll U_j$, z_j becomes linear in x_j, i.e. the hybrid method approximations HA. L_j and U_j are the lower and upper asymptotes to x_j, and x_j will always lie somewhere in between. We will now make approximations in these new variables. The approximations are linear in:

$$y_j \text{ if } \frac{\delta g_i}{\delta x_j} \leq 0$$
$$z_j \text{ if } \frac{\delta g_i}{\delta x_j} > 0$$

L_j and U_j will successively be adjusted in order to improve the approximations. If the subproblem solutions are oscillating in a variable x_j, then L_j and U_j will be forced closer together. On the other hand if the variable x_j is monotonous, i.e. increasing or decreasing, L_j and U_j will be pushed further apart. OASIS uses:

$$L_j^{(k)} = x_j^{(k)} - (\bar{x}_j - \underline{x}_j)$$
$$U_j^{(k)} = x_j^{(k)} + (\bar{x}_j - \underline{x}_j) \quad \text{for } k=1,2$$

and

$$L_j^{(k)} = x_j^{(k)} - s \cdot (x_j^{(k-1)} - L_j^{(k-1)})$$
$$U_j^{(k)} = x_j^{(k)} + s \cdot (U_j^{(k-1)} - x_j^{(k-1)}) \quad \text{for } k>2$$

where
$\quad s = 0.7 \quad$ if $A \leq 0$, i.e. oscillating variables
$\quad s = 1/\sqrt{0.7} \quad$ if $A > 0$, i.e. monotonous variables

and
$$A = (x_j^{(k)} - x_j^{(k-1)}) \cdot (x_j^{(k-1)} - x_j^{(k-2)})$$

We can write the approximated subproblem $\tilde{P}^{(k)}$:

$$\tilde{P}^{(k)}: \quad \min \tilde{w}(x) = w^{(k)} - \sum_{-} (\frac{\delta w}{\delta x_j})^{(k)} (x_j^{(k)} - L_j^{(k)})^2 \left[\frac{1}{(x_j - L_j^{(k)})} + \right.$$

$$\left. - \frac{1}{(x_j^{(k)} - L_j^{(k)})} \right] + \sum_{+} (\frac{\delta w}{\delta x_j})^{(k)} (U_j^{(k)} - x_j^{(k)})^2 \left[\frac{1}{(U_j^{(k)} - x_j)} - \frac{1}{(U_j^{(k)} - x_j^{(k)})} \right]$$

subject to:

$$\tilde{g}(x) = g_i^{(k)} - \sum_{-} (\frac{\delta g_i}{\delta x_j})^{(k)} (x_j^{(k)} - L_j^{(k)})^2 \left[\frac{1}{(x_j - L_j^{(k)})} - \frac{1}{(x_j^{(k)} - L_j^{(k)})} \right] +$$

$$+ \sum_{+} (\frac{\delta g_i}{\delta x_j})^{(k)} (U_j^{(k)} - x_j^{(k)})^2 \left[\frac{1}{(U_j^{(k)} - x_j)} - \frac{1}{(U_j^{(k)} - x_j^{(k)})} \right] \leq \bar{g}_i \quad ; i = 1, I$$

where $\underline{x}_j \leq x_j \leq \bar{x}_j \quad ; j = 1, J$

$\sum_{-}$ and $\sum_{+}$ mean that the summation is over those contributions where the derivatives with respect to x_j are negative and positive respectively. The approximated subproblem $\tilde{P}^{(k)}$ will always be convex, so *dual methods* can be applied. We introduce the Lagrangian function:

$$L(x, \lambda) = \tilde{w}(x) + \sum_i \lambda_i \cdot (\tilde{g}_i(x) - \bar{g}_i), \quad \lambda_i \geq 0$$

where λ_i are the Lagrangian multipliers. L is separable which implies that:

$$L(x, \lambda) = \sum_j L_j(x_j, \lambda)$$

We get $\min_x L(x, \lambda)$ from: $\quad \dfrac{\delta L_j(x_j, \lambda)}{\delta x_j} = 0 \Rightarrow x_j = x_j(\lambda) \; ; \; j = 1, J$

The resulting equations are uncoupled, due to the separability, and of second degree which implies that x_j can be expressed explicitly by the Lagrangian multipliers. Let $\phi(\lambda) = L(x(\lambda), \lambda)$. It can be shown that $\phi(\lambda)$ is concave. λ is found from the solution of the dual problem D:

$$D: \max_\lambda \phi(\lambda)$$

subject to:

$$\lambda_j \geq 0$$

D can be solved by an arbitrary method for unconstrained maximization. OASIS uses the quasi-Newton method. The method is modified to handle the simple non-negative constraints on λ. It should be noticed that at optimum the value of the dual function is equal to the value of the objective function, i.e.:

$$\max \phi(\lambda) = \min \widetilde{w}(x)$$

$\min_x L(x, \lambda)$ implies that:

$$\nabla \widetilde{w}(x) + \sum_i \lambda_i \nabla \widetilde{g}_i(x) = 0$$

which is the Kuhn–Tucker condition for optimum.

5.5 Relaxation. The convexified problem can sometimes be too conservative, which results in a problem with no feasible domain. This situation is avoided by introducing *relaxation* factors h_i for each constraint, eg.:

$$\min \quad \widetilde{w} + \sum_i a_i(h_i + h_i^2)$$

$$\text{subject to:} \quad \widetilde{g}_i \leq \overline{g}_i + h_i \; ; \; i = 1, I$$

$$\underline{x}_j \leq x_j \leq \overline{x}_j \; ; \; j = 1, J$$

$\widetilde{w}$ and $\widetilde{g}_i$ are the approximations of w and g_i respectively, and a_i are constants. Relaxation is automatically included in OASIS–ALADDIN.

5.6 Discrete optimization – the degedewe method. The following method (Esping 1986b) can be applied for variables that are almost continuous. This is often the case for thicknesses, cross section areas and the orientation of fibres in composite materials. For clarity, let us use linear approximations of $\widetilde{w}$ and $\widetilde{g}$ in $\widetilde{P}^{(k)}$ in this illustration. The procedure is however valid for any first order approximation (we normally use the MMA approximations). The user may choose the preferred approximation method (LA/IA or MMA). The procedure is:

Step 1: Continuous optimization starting from $x^{(k)}$ which is the approximation point for subproblem $\widetilde{P}^{(k)}$. The result will be a new continuous design point x^* (fig. 5.2).

Step 2: Discrete optimization based on the same approximation as in 1. The starting point will be x^*. The result of the discrete optimization of subproblem $\widetilde{P}^{(k)}$ will be the new discrete design point $x^{(k+1)}$. The available design space is:

$$x_j \in X_j \qquad X_j = \{x_{1j}, x_{2j}, x_{3j}, ...\}; j = 1, J$$

The values in the design set may be equally or non-equally spaced. It is however a condition that the series of values is monotonous. The subproblem will be:

$$\widetilde{P}^{(k)}: \qquad \min \widetilde{w}(x) = \sum_j (\frac{\delta w}{\delta x_j})^{(k)} (x_j - x_j^{(k)})$$

$$\text{subject to:} \quad \widetilde{g}_i(x) = g_i^{(k)} + \sum_j (\frac{\delta g_i}{\delta x_j})^{(k)} (x_j - x_j^{(k)}) \leq \bar{g}_i = 1$$

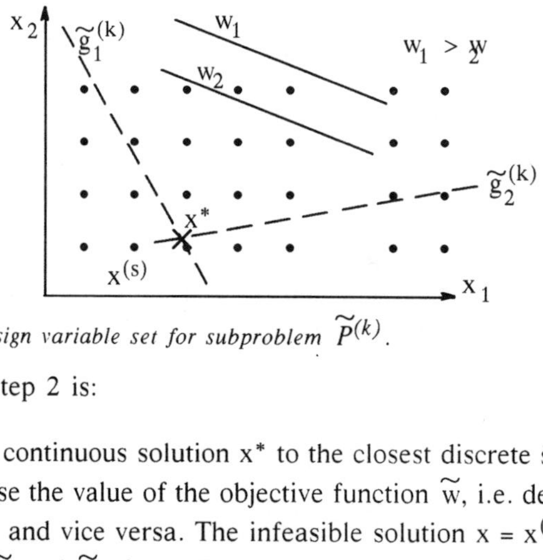

Fig 5.2 Discrete design variable set for subproblem $\widetilde{P}^{(k)}$.

The strategy for step 2 is:

2a: Round off the continuous solution x^* to the closest discrete solution for each x_j in order to decrease the value of the objective function $\widetilde{w}$, i.e. decrease x_j if $\delta w/\delta x_j \geq 0$ for $j = 1, n$, and vice versa. The infeasible solution $x = x^{(s)}$ (fig. 5.2) is then found. Compute $\widetilde{w}$ and $\widetilde{g}_i$ for x. Denote the most critical constraint by $G(x)$, i.e. $G(x) = \max(\widetilde{g}_i)$; $i = 1, I$. Since $\widetilde{P}^{(k)}$ is a first order expression (in this case linear in x) it will be very inexpensive to evaluate $\widetilde{w}$ and $\widetilde{g}_i$ if variable x_j is changed by a single discrete step Δx_j.

2b: Take a step Δx_j for variable x_j to its next value in its set in order to increase the value of $\widetilde{w}$ (the only possible direction to arrive in the feasible domain), i.e:

increase x_j if $\delta w/\delta x_j \geq 0 \Rightarrow \Delta x_j > 0$
decrease x_j if $\delta w/\delta x_j < 0 \Rightarrow \Delta x_j < 0$

Evaluate $\tilde{w}(x_j + \Delta x_j)$ and $\tilde{g}_i(x_j + \Delta x_j)$ and find $G(x_j + \Delta x_j) = \max(g_i(x_j + \Delta x_j))$; $i = 1, I$. Notice that $G(x_j + \Delta x_j)$ probably originates from another constraint $\tilde{g}_i$ than $G(x_j)$. Our objective is to satisfy all $\tilde{g}_i \leq 1$, i.e. $G(x_j + \Delta x_j) \leq 1$. If now $G(x_j + \Delta x_j) < 1$ it will be set to $G(x_j + \Delta x_j) = 1$ in order to avoid an "overfeasible" solution.

Let: $\Delta w_j = \tilde{w}(x_j + \Delta x_j) - \tilde{w}(x_j)$
$\Delta G_j = G(x_j + \Delta x_j) - G(x_j)$

Notice that $\Delta w_j > 0$. If $\Delta w_j = 0$, it will be given a small positive value ϵ, $\Delta w_j = \epsilon$.

We now have two cases:
1. $\Delta G_j \geq 0 \Rightarrow$ the step in x_j does not improve the design. This step will be disqualified.

2. $\Delta G_j < 0 \Rightarrow$ determine $Q_j = \Delta G_j / \Delta w_j$ $(Q_j < 0)$

Q_j will then be determined for all x_j ; $j = 1,J$. The step Δx_k that results in the most negative value of Q_j will then be used to update x, i.e.:
$x_j := x_j + \Delta x_j$ if $j = k$
$x_j := x_j$ if $j \neq k$

2c: If the new design x is feasible $\Rightarrow$ go to E.

2d: Repeat 2b and 2c until a feasible design is achieved. The number of repetitions, r, has an upper limit R, i.e. $r \leq R$, where R is equal to the number of design variables J multiplied by a factor f. We have used f = 3.

2e: Stop.

There is risk that there are no feasible discrete solutions to the subproblem. In that case the constraints $\tilde{g}_i$ will be relaxed by a factor $\underline{G}$ where $\underline{G}$ is the minimum value of G(x) that ever occurred in 2a or 2b. The procedure 2a–2e will then be repeated once more.

6. Examples

The following real life examples include applications from the industry, such as turbine machinery, automotive, sail yacht, rolling stock, and industrial tool components. They embrace aluminum castings, steel forgings, sheet steel stampings, and GRP sandwich; and examplify objective functions of stress, weight, stiffness, and natural frequency; with constraints on weight, contact forces, stiffness, deflections, stresses, as well as explicit geometric constraints. They are taken from our experience as consultants to the industry, by the kind permission of our clients.

6.1 Minimization of the maximum stress. In this case the objective was to minimize the maximum occurring effective stress at the interface between a gas turbine blade and the hub. A severe complication in the problem was the behaviour of the contact zone, which had to be taken into account. By formulating the contact problem as strain constraints in special elements introduced in the contact zone, the contact problem could be included in the actual optimization and solved correctly for every iteration, even as the shape of the contact zone and the parts changed. As a result a new shape was found such that the maximum effective stress was reduced by one order of magnitude.

6.2 Weight minimization. In collaboration with Saab–Scania AB, Saab Car Division, a design study was performed aimed at improving the rear suspension of a SAAB 9000 car by minimizing the weight of the suspension arm. Five shape variables and one thickness variable were introduced, and constraints were formulated to prevent the stress levels in the optimized arm from exceeding those in the original design. Deformation constraints were applied to the flanges to keep them from twisting into contact with the wheel rims. The optimization reduced the weight (which is unsprung) by 10%. See Esping et al. (1987b).

6.3 Stiffness maximization. The purpose of this project was to maximize the bending stiffness of the GRP–sandwich hull of an R12m yacht, at a given load. The weight was fixed by the rules of the International Twelve Metre Association (ITMA); the most profitable objective was therefore to maximize the hull stiffness. Increased hull stiffness means improved performance and speed. First a single web-to-web section of the hull was optimized subjected to slamming loads. The weight of the section was minimized with constraints on panel deflections. The resulting thicknesses were then used as lower limits in the optimization of the complete hull. The complete hull was then subjected to rig loads. The longitudinal bending stiffness was maximized with constraints on the weight and the deflections. The weights per unit area were not allowed to be less than those proposed by the ITMA, and the total weight was not allowed to be higher than that of a conventional aluminum hull. Constraints on panel- and torsional stiffness were also applied. The hull was divided into nine different parts, each with an unique sandwich lay-up, in accordance with the rules proposed by the ITMA. The model included 79 variables and

34 constraints. The optimization redistributed the weight of the sandwich hull so that the stiffness was increased by more than 58% as compared to the original design. Compared to a conventional aluminum hull of equal weight, the stiffness was increased by 30%. See Ljunggren (1990).

6.4 Maximized natural frequency. The objective in this case was to establish whether a GRP sandwich concept was suitable for use in a subway car. The potential advantages would be a lower structural weight and lower propulsion costs, better acoustics and thermal insulation. The decisive design criterion was the natural frequency. The lowest eigenfrequency of the longitudinal bending mode of the structure had to be at least 8 Hz, since vibrations may otherwise cause considerable discomfort for the passengers. Design variables were attached to the thickness of the core and the thicknesses of both faces, in a number of different areas of the structure. A total of 63 design variables were active in the optimization. As a result of the optimization the value of the lowest eigenfrequency was increased by 70% (from 4.7Hz to 8.0Hz), at the moderate cost of a weight increase of 15%. The final weight was still below the weight of the original aluminum design by a good margin. See Isby (1990).

6.5 Increased "toughness". The shape of the cast aluminum handle of a pneumatic riveting hammer from Atlas Copco Tools was optimized with the aim to improve its resilience and ability to withstand rough treatment, especially accidental loads such as those experienced when falling onto a hard floor from a high scaffolding or being run over by a truck, and to quantify the improvement. Three load cases were included. For one case the "toughness" was increased by more than 100%, for the other two by lesser but significant amounts. Tool performance was not impaired in any way, in fact the already good ergonomic design was improved in the process.

7. Summary

Structural optimization has reached the point where it is actively and successfully used in the engineering industries, and the results are being implemented in the final design and production of actual components, not only prototypes and test cases. Depending on its size and resources, a company may choose either to acquire optimization software themselves and develop optimization experience and competence in-house, or to purchase specific optimization services from a firm specializing in the field. For specialized applications it is also possible to have dedicated and customized software developed, both quickly and at a low cost. Furthermore, it is possible for any company possessing in-house code for specific analyses to integrate their programs into a multi-disciplinary package for product optimization.

OASIS-ALADDIN is an integrated and self contained system for large scale structural optimization in the linear response domain. It is used today in diverse applications in aerospace, automotive, marine, defense and construction industry, among others.

OCTOPUS is a software tool for concurrent, integrated multi-disciplinary optimization where objective and constraint functions from many different disciplines can be combined and treated simultaneously. The system has an open architecture where user written programs can be easily integrated.

References

Esping, B. (1984), Minimum Weight Design of Membrane Structures using Eight-node Isoparametric Elements and Numerical Derivatives, Computers & Structures, vol 19, no 4.
Esping, B. (1985a), Analytical Derivatives of Structural Matrices for an Eight-node Isoparametric Membrane Element With Respect to a Set of Design Variables, Computers & Structures, vol 21, no 3.
Esping, B. (1985b), A CAD Approach to the Minimum Weight Design Problem, Int Journal for Numerical Methods in Engineering, vol 21.
Esping, B. (1986a), The OASIS Structural Optimization System, Computers & Structures, vol 23, no 3.
Esping, B. (1986b), The ALIBABA Nonlinear Optimization Package, Computers & Structures, vol 24, no 2.
Ringertz, U., Esping, B. and Bäcklund, J. (1986), Computer Sizing of Sandwich Constructions, Composite Structures 5.
Svanberg, K. (1987), Method of Moving Asymptotes – A New Method for Structural Optimization, Int Journal for Numerical Methods in Engineering, Vol. 24, 359-373.
Esping, B. and Holm, D. (1987a), A CAD Approach to Structural Optimization, NATO ASI Series, Vol. F27, Computer Aided Optimal Design: Structural and Mechanical Systems, edited by C.A. Mota Soares. Springer-Verlag, Berlin, Heidelberg.
Esping, B., Holm, D., Isby, R. and Larsson, M. (1987b), Shape Optimization of a Suspension Arm Using OASIS, Int Journal of Vehicle Design.
Esping, B., Holm, D., Olsson, K. and Lönnqvist, J. (1987c), Shape Optimization of a Truck Front Axle Beam, Report 87-7, Dept. of Aeronautical Structures and Materials, The Royal Institute of Technology, Stockholm.

Esping, B. and Holm, D. (1988), Structural Shape Optimization using OASIS, in Structural Optimization, edited by Rozvany, G.I.N. and Karihaloo, B.L., Kluwer Academic Publishers, Dordrecht.

Kim, K.J., Esping, B. and Holm, D. (1989), A Numerical Method for Minimizing Ship Resistance, SSPA Report No 2964-1, Gothenburg.

Ljunggren, L. (1990), Strukturoptimering av skrov i glasfibersandwich till segelbåt typ R-12m, ALFGAM Optimering AB rapport 90-01-01, Stockholm, in Swedish (Structural Optimization of a GRP Sandwich Hull for an R-12m type Sailing Yacht).

Samuelsson, J., Holm, D. and Esping, B. (1990), Optimization of Hydraulic Cylinder Housing, Int Journal of Fatigue, 12 No 6.

Isby, R. (1990), Optimization of Railway Car in Sandwich Design – Natural Frequency (Example: Subway Car), Report 90-1, Dept. of Aeronautical Structures and Materials, The Royal Institute of Technology, Stockholm.

OASIS-ALADDIN User's Manual, 3rd ed. (1991), ALFGAM Optimering AB, Stockholm.

Gullberg, O. and Romell, O. (1991), Structural Optimization of a High Performance GRP-Sandwich Ship Hull, FAST'91, proceedings of the First International Conference on Fast Sea Transportation, Trondheim, Norway, 17-21 June.

Svanberg, K. (1991), MMA With Some Extensions, presented at the NATO Advanced Study Institute conference Optimization of Large Structural Systems, Berchtesgaden, September 23-October 4.

Esping, B., Clarin, P. and Romell, O. (1991), OCTOPUS – A Tool for Distributed Optimization of Multi-Disciplinary Objectives, presented at the NATO ASI Optimization of Large Structural Systems, Berchtesgaden, Germany, 23 September-4 October.

Romell, O., Esping, B., Holm, D., Clarin, P. and Ljunggren, L. (1991), Structural Optimization using the OASIS-ALADDIN System, presented at StruCoMe'91, Paris, 18-21 November.

Author's address:

ALFGAM Optimering AB
Roslagsvägen 101, hus 15
S-104 05 STOCKHOLM, Sweden
Phone: +46-8-790 98 66
Fax: +46-8-15 97 17
E-mail: alfgam@ilk.kth.se

The Structural Optimization System OPTSYS

Torsten Bråmå

Abstract. OPTSYS is a modular structural optimization system with well defined interfaces to FE–programs and codes for aeroelasticity. A mathematical programming approach is adopted were a sequence of convex approximations of the initial problem is solved, using the MMA method. This approach makes it possible to take all design criteria into account simultaneously. Gradients are calculated semi–analytically. OPTSYS can treat design variables associated to the shape of the structure, the element cross section properties or the material direction in the case of composite materials. Design criteria can be defined on weight, moment of inertia, displacement, stress, eigenfrequency, local buckling, flutter and aileron efficiency. Recent developments has concerned dynamic response and acoustics. Other important ingredients are; the integration of a preprocessor to define shape variables, the treatment of discrete variables and the possibility to deal with substructured FE models. The capabilities and methods will be discussed and illustrated with applications on aircraft and automotive structures.

1. Introduction

This paper describes the capabilities of OPTSYS version 3.0, released in 1992. The program originates from an early version of the OASIS system developed by Esping (1986) and has since 1984 been developed further at Saab Aircraft Division together with the Aeronautical Research Institute in Stockholm. A major contribution has also been made by Svanberg (1987) at the Royal Institute of Technology.

The OPTSYS system will first be briefly described. Next the possibilities in problem formulation and the methods for the solution of the optimization problem are presented.

OPTSYS has been applied to both aerospace and automotive structures, e.g. an investigation of the potential weight savings in a composite wing of a fighter aircraft involving more than 700 design variables, simultaneous shape and thickness optimization of a Saab 9000 car suspension arm. These applications have been presented in Bråmå (1990) but included here for the sake of completeness, followed by an application concerning reduction of cabin noise in a civil aircraft.

2. Software

Figure 1 shows a simplified picture of the system in terms of included software and files. The pre–processor can be used to create both input to FE–programs and at least parts of the OPTSYS–input (primarily the linking between the formulation of the optimization problem and quantities in the FE model).

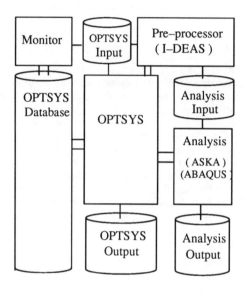

Figure 1

In the case of shape optimization the pre–processor is also executed in batch mode in order to collect updated node coordinates corresponding to current shape variable values. The interface consists of files with documented format. The standard pre–processor is now I–DEAS (previously PREFEM) but any other software, able to read and write the interface files, may be used.

The analysis programs involved are two alternative FE–programs (ASKA or ABAQUS) and codes for aeroelastic analysis (AEREL and WINGBODY).

The sensitivity analysis is performed inside OPTSYS collecting the required data from the analysis programs.

OPTSYS consists of a number of modules communicating through the OPTSYS database. The sequence of modules to be executed depends of the application. The database is an application of MEM–COM described in Merazzi (1991).

The Monitor is a collection of pre– and post– processing functions for problem formulation, diagnosis and documentation of the optimization process.

The iteration process, illustrated in figure 2, can briefly be described as follows. First the current design is analyzed with respect to all required design criteria. The different analyses are often independent but, for instance, the aeroelastic programs require structural stiffness data from a preceding FE analysis.

Next the status of the optimization process is evaluated and an active set strategy is used to select which gradients to calculate. The gradients are then calculated with the same dependencies as in the analysis step.

Finally in the redesign step an explicit subproblem is formulated and solved producing a new set of design variables.

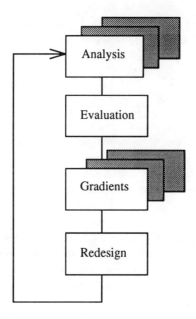

Figure 2

3. Theory

List of symbols:

K, D, M	stiffness, damping and mass matrix,
ω	eigenfrequency
Φ	eigenmode
ω_F	excitation frequency
u	response vector
F	load vector
k_e, d_e, m_e	element matrices
v_e, u_e	element parts of the response vectors
X	design variable vector
x	design variable
ξ	element variable
w(**X**)	objective function
g(**X**)	constraint function
f(**X**)	primary function

3.1 Formulation and solution of the optimization problem.

A mathematical programming approach is adopted with the following general problem formulation.

$$\text{Min } w(X)$$
$$g_i(X) < 1$$
$$X_{min} < X < X_{max}$$

The design variables **X** are linked to element properties ξ in the FE model ; cross section area in rod elements, thickness in shell or membrane elements, stiffness in spring elements, material direction in shell/membrane elements, individual node coordinates or general shape of the structure defined through a geometry model in a preprocessor. For instance can the thickness of several elements be linked to the same design variable by the linear relation, $\xi(x)$.

In the case of composite material, the composite stack is defined in OPTSYS so that each layer corresponds to all layers of the same type (e.g. 0, 90 +45, –45 degrees). One or several elements are associated to the same stack and design variables are linked to the layer thicknesses. The design variables can be discrete, i.e. correspond to a number of layers.

The optimization problem is built up by primary functions f(**X**). Available function types are:

- weight or moment of inertia,
- general displacement,
- stress, strain or local buckling,
- spring forces,
- structural or acoustic dynamic response,
- eigenfrequency or eigenmode,
- flutter damping,
- aeroelastic efficiency criteria.

In addition, combination functions $F(f(X),X)$ can be built as a general polynomial in $f(X)$ and X.

The objective function $w(X)$ can be equal to one selected function $f(X)$ or $F(X)$. Another possibility is to define $w(X)$ as the maximum within a set of normalized function values and thus formulate a Min(Max) problem.

The constraint functions $g(X)$ are defined by combining a maximum or minimum limit with a function $f(X)$ or $F(X)$.

In each global iteration an explicit convex subproblem is formulated using the MMA method described in Svanberg (1987), where first order gradients of the functions are needed. If discrete variables are included in the problem, MMA makes a search for the best feasible discrete point in the neighborhood of the continuous solution.

3.2 Sensitivity analysis. A semi–analytical approach is adopted where the derivatives of element properties are derived numerically.

$$\frac{\partial k_e}{\partial \xi} = \frac{k_e(\xi+\Delta\xi) - k_e(\xi)}{\Delta\xi}$$

Element properties considered so far are weight, moment of inertia, stiffness matrix, mass matrix, strain–displacement matrix, load vector and transformation matrix to material directions.

The increments are derived differently depending on the type of design variable. For example, if it is thickness variable the increment is simply a constant factor times the current thickness. If it is a shape variable, disturbed node coordinates are first produced by the preprocessor corresponding to increments in the shape variable values. The disturbed node coordinates are then used when calculating the element properties.

The element properties and their derivatives are stored on the database and used later in different types of sensitivity calculations.

Static response functions

The static response problem in matrix form ;

$$\mathbf{K}\,\mathbf{u} = \mathbf{F}$$

The displacement function is generally formulated as a linear combination of the components in the response vector.

$$f = \mathbf{q}^t\,\mathbf{u}$$

The derivative of f with respect to a design variable x is derived as follows, assuming that **q** not is dependent of x ;

$$\frac{\partial f}{\partial x} = \mathbf{q}^t\,\frac{\partial \mathbf{u}}{\partial x} = \mathbf{v}^t\,\frac{\partial \mathbf{F}}{\partial x} - \mathbf{v}^t\,\frac{\partial \mathbf{K}}{\partial x}\,\mathbf{u}$$

where **v** is the solution to the following problem ;

$$\mathbf{K}\,\mathbf{v} = \mathbf{q}$$

An element strain component is also a linear combination of the components in the response vector and can be calculated similarly.

$$\varepsilon = \mathbf{q}^t\,\mathbf{u}$$

The vector **q** corresponds in this case to the element strain–displacement relation which depends on the element formulation. In this case **q** may be dependent of x ;

$$\frac{\partial \varepsilon}{\partial x} = \mathbf{v}^t\,\frac{\partial \mathbf{F}}{\partial x} - \mathbf{v}^t\,\frac{\partial \mathbf{K}}{\partial x}\,\mathbf{u} + \mathbf{u}^t\,\frac{\partial \mathbf{q}}{\partial x}$$

In the case of anisotropic material, it is desired to calculate strains in the material direction. If the material direction is connected to a design variable we get the derivative as ;

$$\frac{\partial \varepsilon_m}{\partial x} = \frac{\partial \mathbf{T}_m}{\partial x}\,\varepsilon + \mathbf{T}_m\,\frac{\partial \varepsilon}{\partial x}$$

$\mathbf{T}_m$ is here the transformation matrix to material directions.

Derivatives of corresponding stresses can now be calculated as ;

$$\frac{\partial \sigma}{\partial x} = E \frac{\partial \varepsilon}{\partial x}$$

assuming here that Hooks matrix **E** is constant.

The local buckling criteria in bar elements is formulated as

$$\frac{\sigma L^2}{E c a} < 1$$

where
- σ the stress in the bar (compression positive)
- L   the length of the bar
- a   the area of the bar
- E   the module of elasticity
- c   the buckling constant which depends on the cross section shape (not the size !) and the boundary conditions.

Applying the chain rule again we get

$$\frac{\partial}{\partial x}\left(\frac{\sigma L^2}{E c a}\right) = \frac{L^2}{E c a}\frac{\partial \sigma}{\partial x} - \frac{\sigma L^2}{E c a^2}\frac{\partial a}{\partial x} + \frac{\sigma 2 L}{E c a}\frac{\partial L}{\partial x}$$

The local panel buckling criteria has the following form ;

RG = FUNC(R1, R2, R12) ,

where R1, R2 and R12 are the ratios between current stress and critical buckling stress for each stress component. RG takes the total stress state into account by applying the function FUNC to the individual components. FUNC is evaluated from experimental data and explicitly stored into the software.

Eigenfrequency functions

The structural eigenvalue problem in matrix form ;

$$(K - \omega^2 M)\Phi = 0$$

The derivative of one eigenfrequency with respect to a design variable x is then calculated according to;

$$\frac{\partial \omega^2}{\partial x} = \frac{1}{\Phi^t M \Phi} \Phi^t \left(\frac{\partial K}{\partial x} - \omega^2 \frac{\partial M}{\partial x} \right) \Phi$$

assuming that the eigenmode is normalized with respect to M, the derivative can be calculated with contributions from affected elements as ;

$$\frac{\partial \omega^2}{\partial x} = \sum_e \left[\Phi_e^t \left(\frac{\partial k_e}{\partial x} - \omega^2 \frac{\partial m_e}{\partial x} \right) \Phi_e \right]$$

Dynamic response functions

The treatment of dynamic response function is similar to the static case but the character of the dynamic response function is however not as attractive to deal with. The dynamic response is not a monotonous function of structural size variables, as a maximum will occur when an eigenfrequency gets close to the excitation frequency. This will lead to unconnected feasible regions in the design space which is a major difficulty for the optimization algorithm.

The dynamic response problem in matrix form ;

$$(K + i \omega_F D - \omega_F^2 M) u = F$$

The vectors **u** and **F** are complex and represent the amplitude and phase in the harmonic vibration. The response vector consists in the structural parts of the usual displacements depending on the type of finite element used and in the acoustic cavity parts of the model we find the acoustic pressure as one degree of freedom.

The dynamic response function is now defined as the absolute value of a linear combination of components in the response vector defined as

$$f = q^t u$$

q is a vector containing combination coefficients and

$$|f| = \sqrt{f_{Re}^2 + f_{Im}^2}$$

where Re indicates the real part and Im the imaginary part.

In the acoustic case the function can alternatively be defined as the sound pressure level, SPL, defined as;

$$SPL = 20 \log \frac{|f|}{\sqrt{2}\, p_o}$$

where p_o is the acoustic reference pressure.

Using the symbol Q for the system matrix

$$Q\,u = F$$

Differentiating with respect to one variable x and pre-multiplying with q^t gives :

$$q^t \frac{\partial u}{\partial x} = v^t \frac{\partial F}{\partial x} - v^t \frac{\partial Q}{\partial x} u$$

where v is the solution of

$$Q\,v = q$$

Here v is complex if Q is complex. Assuming that the derivative of F is zero, we can now write;

$$\frac{\partial f}{\partial x} = q^t \frac{\partial u}{\partial x} = -\sum_e \left[v_e^t \left(\frac{\partial k_e}{\partial x} + i\omega_F \frac{\partial d_e}{\partial x} - \omega_F^2 \frac{\partial m_e}{\partial x} \right) u_e \right]$$

Finally the derivative of the absolute value can be expressed as

$$\frac{\partial |f|}{\partial x} = \frac{f_{Re}}{|f|} \frac{\partial f_{Re}}{\partial x} + \frac{f_{Im}}{|f|} \frac{\partial f_{Im}}{\partial x}$$

and the derivative of the sound pressure level as

$$\frac{\partial (SPL)}{\partial x} = \frac{20 \log e}{f} \cdot \frac{\partial f}{\partial x}$$

Flutter damping

Flutter is a serious vibration phenomenon which, if it occurs, might be disastrous. It is therefore vital to be able to avoid flutter in the structural design process. The AEREL system described in Stark (1990) is used for flutter analysis. The analysis of this aerodynamic instability yields a nonlinear and complex eigenvalue problem. The location of the eigenvalues in the complex plane indicate if the vibrations are stable or not.

AEREL first calculates generalized aerodynamic forces (transfer functions) using separate AEREL modules for subsonic and transsonic speed. Then the nonlinear (and complex) eigenvalue problem is solved.

$$\left(K_0 + \omega D_0 + \omega^2 M_0 + \frac{v^2}{\pi \mu} A\left(\frac{\omega}{v}\right)\right) \Psi = 0$$

where

$K_0 = K / (m_r \omega_r^2)$	dimensionless stiffness matrix
$D_0 = D / (m_r \omega_r)$	dimensionless damping matrix
$M_0 = M / m_r$	dimensionless mass matrix
$A(p) = A(\omega / v)$	aerodynamic transfer function
ω	flutter eigenvalue
Ψ	flutter eigenmode
m_r	reference mass
ω_r	reference frequency
S, L	reference area and length
U, ρ	free-stream speed and density
$v = U / (\omega_r L)$	dimensionless free stream speed
$\mu = 2 m_r / (\pi \rho S L)$	mass ratio

K, D, M and **A** are expressed in a base of m selected structural eigenmodes, Φ. The eigenmodes and the matricies **K** and **M** are obtained from the FE model calculations or from ground vibration tests. D_0 can likewise be obtained from tests but can usually be neglected. Matrix **A**, which is calculated by AEREL, depends on the Mach number and the Laplace transform parameter $p = \omega / v$. ρ depends on the air density (altitude).

The complex eigenvalues indicate if the modes are stable or not. If the damping factor defined by

$$g = -\frac{\omega_{Re}}{\omega_{Im}}$$

is negative, the associated mode is unstable. The imaginary part is the circular frequency of the flutter mode.

For desired combinations of Mach numbers and altitudes, m eigenvalues are calculated.

The flutter constraint is formulated to assure a certain amount of damping for all modes and for all flight cases specified.

Using the symbol $\mathbf{Q}$ for the system matrix, the eigenvalue problem is written

$$\mathbf{Q}\Psi = 0$$

and we define the associated vector Ψ_a by

$$\mathbf{Q}^t \Psi_a = 0$$

The vectors Ψ and Ψ_a are not identical since $\mathbf{Q}$ is not symmetric.

By differentiating with respect to a design variable x and multiplying by Ψ_a^t, we get

$$\Psi_a^t \frac{\partial \mathbf{Q}}{\partial x} \Psi = 0$$

and then

$$\frac{\partial \omega}{\partial x} = -\frac{1}{c} \Psi_a^t \left(\frac{\partial \mathbf{K}}{\partial x}0 + \omega \frac{\partial \mathbf{D}}{\partial x}0 + \omega^2 \frac{\partial \mathbf{M}}{\partial x}0 \right) \Psi$$

where

$$c = \Psi_a^t \left(2\omega \mathbf{M}_0 + \mathbf{D}_0 + \frac{v}{\pi \mu} \frac{\partial \mathbf{A}}{\partial p} \right) \Psi$$

Differentiating the damping factor with respect to one variable x, we immediately get

$$\frac{\partial g}{\partial x} = \frac{1}{\omega_{Im}^2} \left(\omega_{Re} \frac{\partial \omega_{Im}}{\partial x} - \omega_{Im} \frac{\partial \omega_{Re}}{\partial x} \right)$$

The contribution from $\partial \mathbf{D}_o / \partial x$ is neglected in the program, while $\partial \mathbf{A} / \partial p$ is calculated in AEREL. Accurate calculation of the aerodynamic transfer functions is time consuming and such a calculation is therefore done in AEREL only for a limited number of discrete values of p. A linear combination of simple analytic functions is then fitted to the discrete values and employed in the final routine. The additional statements required in AEREL for calculation of $\partial \mathbf{A} / \partial p$ (via differentiation of the combination) is therefore very simple.

$\mathbf{K}$, $\mathbf{M}$, $\partial \mathbf{K} / \partial x$ and $\partial \mathbf{M} / \partial x$ are calculated by OPTSYS using information obtained from the FE analysis.

For instance, the derivative of the stiffness matrix is calculated as a sum over finite elements affected by x;

$$\frac{\partial \mathbf{K}}{\partial x} = \sum_e \Phi_e^t \frac{\partial \mathbf{k}_e}{\partial x} \Phi_e , \quad (m \times m)$$

4. Industrial applications

4.1 Shape optimization of a Saab 9000 suspension arm. In order to investigate the performance of a proposed new wishbone design (figure 3) for the Saab 9000 car, an optimization project was initiated. The new design is of forged aluminium, the one in production is built from pressed steel parts. Optimization is important here since a low unsprung weight of the suspension is crucial for a performance car. A simple problem formulation for a first re–design attempt was sought.

A FE–model consisting of 230 shell elements was applied with three loading cases; maximum straight line breaking, maximum lateral acceleration (cornering) and maximum combined braking/lateral acceleration.

The cross sectional properties along the wishbone was varied by having the thickness of the elements as variables in the optimization problem. The inner boundary was described by B–splines in the geometry description of the preprocessor PREFEM (1987). The control–points of these splines were connected to design variables. Upper and lower limits on the values of the design variables accounted for various geometrical limitations (figure 4).

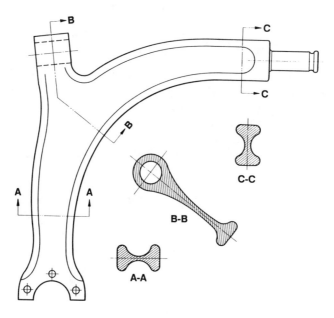

Figure 3. Wishbone layout

Stress constraints were defined to keep the maximum von Mises stress below the yield stress. The basic stiffness requirement was that the stiffness of of the new wishbone should equal the stiffness of the original (steel).

The resulting optimization problem contained a total of 122 thickness variables, 6 shape variables, 1300 stress and 6 deflection constraints.

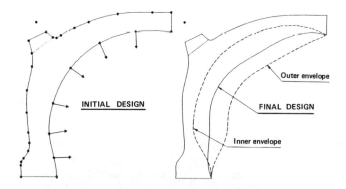

Figure 4. Geometry of initial and final design

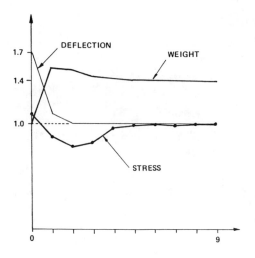

Figure 5. Iteration history

The problem was solved in 9 iterations. For a weight increase of 40 percent OPTSYS found an optimal solution with sufficient stiffness (63 percent increase). The final design was determined, for this problem statement, completely by the stiffness requirements, two of which were at the critical limit. The stress constraints had no impact on the final design as they all were non–critical (albeit very close). Results are shown in figures 4, 5 and 6.

The thickness distribution of the final design was dominated by the defined lower limit. The exception being the far "left" part which thickness probably was increased to create enough stiffness for the lateral load.

The average CPU time per iteration, on a VAX 8800, was roughly 550 seconds including the FE analysis part taking about 100 seconds.

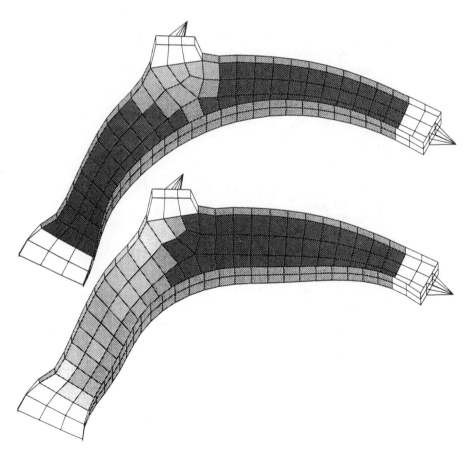

Figure 6. Thickness of initial and final design. Dark – thin. Light – thick.

4.2 Composite wing of the Gripen aircraft.

The main purpose of this very large application was to investigate the possible weight savings for redesign of the wing skins with two choices of new composite materials.

A substructured FE–model of the complete aircraft was used. By including the optimizationwise active parts of the wing structure in a separate substructure, the amount of calculations needed in each iteration was reduced to a reasonable size. The active substructure contained about 5000 degrees of freedom compared to the 125000 in the complete aircraft model. Eight loading cases were selected for this study.

The design variables were associated to layers in 254 different composite stacks. The layup in each stack was defined by three independent variables controlling the number of 0 degree layers, 90 degree layers and +/– 45 degree layers, making a total of 762 design variables. One or several finite elements in the wing panels were then linked to each stack. Explicit linear constraints were defined on the sum of all thickness variables connected to the same stack to limit the total thickness of the wing panel. Constraints were also imposed on fibre strain and local buckling in the composite. Constraints on the aircraft performance such as aeroelastic efficiency should ideally also have been included. However, as the criteria was to maintain current performance, it was here considered sufficient to formulate the aeroelastic requirements as a number of constraints on the wing torsion. A total of about 20000 potential constraints were defined of which a few hundred were active in the final design.

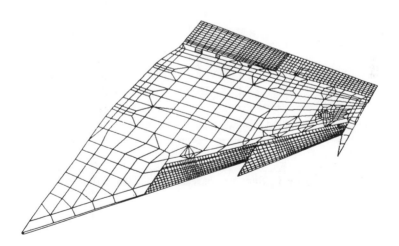

Figure 7. FE–model of wing substructure

Six global iterations were enough to solve this problem for each of the two alternative materials. Each iteration needed approximately 2000 CPU seconds in the CRAY 1–A ; 130 seconds for the reanalysis, 1000 seconds for the gradient calculation and 800 seconds for the solution of the approximate subproblem. The portion of the iteration time consumed by the subproblem solution was much larger here than in smaller problems. One way to reduce this portion is to lower the accuracy in the solution of the subproblem.

The layups produced by OPTSYS have to be adjusted to production requirements impossible to account for in the original problem formulation. This manual work leads of course to increased weight and can be very tedious. Good post processing aids are absolutely vital when dealing with the huge amount of information created in large applications like this.

A summary of weight savings and the relative material data can be found in figure 8. The results in terms of optimal layups will be valuable in a possible future redesign of the wing.

		Initial wing	HT material	IM material
Matrix :		Brittle epoxy	Toughend epoxy	Toughend epoxy
Elastic modulus: $E11$		0.97	1.0	1.2
Allowables:	ε_t	0.8	1.0	1.0
	ε_c	0.92	1.0	0.86
Price:		0.9	1.0	1.5
Weight saving:		(Not optimized)	14 %	20 %

Figure 8. Summary of results and material data

4.3 Reducing cabin noise in the Saab 340 aircraft.
Passenger comfort is of great importance in most transport vehicles. For instance, in the new generation of regional turboprop aircraft, a low noise level is vital to be competitive on the market. The possibilities to predict noise levels analytically has improved rapidly in recent years, see Sandberg and Göransson (1988). This will make it possible to take acoustic design criteria into account in early project stages.

The 2-D FE-model (figure 9), representing a cross section of the Saab 340 fuselage close to the plane of the propeller, consists of one substructure for the structural part and another substructure for the cavity. The cavity substructure contains 2-D acoustic elements and interface elements connecting the cavity model to the outer flange. Four tuned dampers are included in the model. The tuned damper is modeled as a point mass connected to the structure with a spring parallel to a dashpot. The introduction of the dashpot makes the system matrix complex. The excitation from the propellers on the outside of the fuselage (amplitude and phase) is expressed as complex nodal forces in the FE-model.

The design variables are chosen to be the cross section area of the inner flange to investigate how much stiffening of the frame can reduce the cabin noise. The objective function is the weight of the inner flange, i.e. the weight of the elements associated to design variables. The acoustic design criteria is an upper limit of the sound level in three points in the cabin corresponding to measurement points in flight tests.
The location of the four tuned dampers, the 37 design variables and the three constraint points are indicated in figure 10.

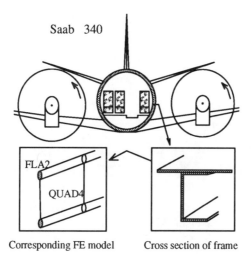

Figure 9

The desired sound level is reach after five iterations by adding material mainly at design variables 24, 25 and 26. The iteration history is given in figure 11. Figure 12 shows the real part of the final structure response including both the structural displacements and the air pressure distribution.

This application was a test problem in connection with the development of acoustic constraints. In a more realistic application of course other design criteria have to be considered as well. Current development includes the possibility to link the properties of the tuned damper to a design variable.

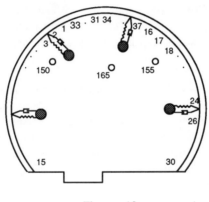

Figure 10

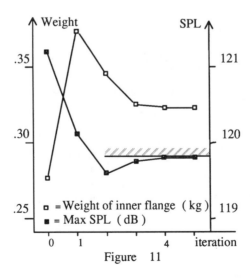

Figure 11

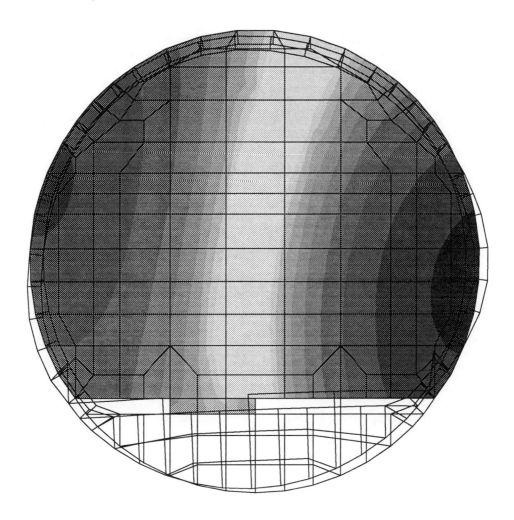

Figure 12. Sound level and real part of structural response

5. Conclusions and future plans

OPTSYS has proved to be able to deal with a number of different types of structural optimization problems. The introduction of this kind of software tools in the design process has however only started. There is a great potential for a much wider use when engineers have learned what can be done and how. It is a great challenge to really make the systems easy to use and well integrated in the CAE environment.

The immediate development plans for OPTSYS includes the introduction of beam elements and a multi–model capability which will allow more than one model of the structure to be treated simultaneously.

A procedure for model updating, described in Larson et al. (1992), is also being developed, where the FE–model is to be adjusted to have improved dynamic properties when compared with experimental data.

References

Esping, B.J.D. (1986), The OASIS structural optimization system, Computers and Structures, vol. 23, No 3, pp. 365–377.

Svanberg, K. (1987), The Method of Moving Asymptotes – a new method for structural optimization, International journal for numerical methods in engineering, Vol. 24, pp. 359–373.

PREFEM (1987), User manual, version 5.4, Veritas Sesam Systems A.S., Norway.

Sandberg, G. and Göransson, P. (1988), A Symmetric Finite Element Formulation for Acoustic Fluid–Structure Interaction Analysis, JSV, Vol. 123 (2).

Bråmå, T. (1990), Applications of the structural optimization program OPTSYS, ICAS–90–2.1.3.

Stark, V. J. (1990), The AEREL Flutter Prediction System, ICAS–90–1.2.3.

Merazzi, S. (1991), MEM–COM USER MANUAL, version 6.0, SMR Corp, Switzerland.

Larsson P.O. et al. (1992), Model updating in a software environment for Structural Optimization, IMAC 10, 3–6 February 1992, San Diego, Cal. USA.

Authors address:

Torsten Bråmå
Saab Military Aircraft
Linköping
Sweden

SAPOP
An Optimization Procedure for Multicriteria Structural Design

Hans A. Eschenauer   Johannes Geilen   H. Jürgen Wahl

Abstract. The present paper describes the optimization procedure SAPOP developed at the Research Center for Multidisciplinary Analysis and Applied Structural Optimization. Following a general introduction, the basic principle of SAPOP is presented in form of the so-called "Three-Columns-Concept". After that, the program-technical realization is described. Finally the efficiency is exemplified by various cases of application from the industrial practice.

1. INTRODUCTION

The goal of structural optimization is to support the engineer in searching for the best possible design alternatives for specific structures. The "best possible" or "optimal" structure here applies to that structure which most corresponds to the designer's desired concept and his objectives, while at the same time meeting operational, manufacturing and application demands. Compared with the "Trial and Error"-method generally used in engineering practice and based on an intuitive empirical approach, the finding of optimal solutions by applying mathematical optimization procedures is more reliable and efficient. These procedures can be expected to be more frequently applied in industrial practice. The following aspects show the necessity of introducing optimization procedures into the practical design phase:

- Increasing the quality and quantity of products and plants while at the same time reducing costs and thereby increasing competitiveness.
- Fulfilling the ever-increasing demands for reliability, safety, pollution control and energy saving
- Introducing inevitable rationalization measures in development and design offices (CAD, CAE) in order to save more time for the staff to work creatively.

During the last years these assumptions have lead to development of the "Three-Columns-Concept" at the Research Center for Multidisciplinary Analysis and Applied Structural Optimization (FOMAAS). The columns are structural model, optimization model and optimization algorithms. This concept has been realized then in the optimization procedure SAPOP (**S**tructural **A**nalysis **P**rogram and **O**ptimization **P**rocedure) as an academic structural optimization system. So far, the following special capabilities have been implemented within SAPOP among others:

- multicriteria design optimization,
- decomposition methods,
- shape and sizing optimization,
- stochastic and topology optimization and
- consideration of composite materials.

The program system employs different structural analysis methods which, on the one hand, are commercial systems (e.g. Finite element analysis such as ANSYS or SAP) and, on the other hand, have been developed by the the users (e.g. analytical methods, transfer matrices, finite difference method). Additionally, the optimization procedure offers efficient commercial pre- and post-processors. Furthermore, various optimization algorithms have been implemented which can be applied depending on the problem. The program system is used in research, teaching and for practical applications in the scope of industrial projects.

2. FUNDAMENTALS

2.1 Notation

Because of the great number of different notations in structural optimization, some of the terms commonly used in the optimization procedure SAPOP shall be defined in alphabetical order:

Analysis variables : structural parameters varied during optimization computations,

Constraints : mathematically formulated design requirements not covered by the objective(s); a constraint is active, if it equals zero,

Design model : mathematical link between the design variables and analysis variables,

Design variables : design quantities to be varied,

Evaluation model : mathematical link between the state variables and the objective function and constraint values under consideration of optimization strategies,

Feasible domain : design space surrounded by the constraints,

Final design : optimal values of the design variables at the end of the optimization process,

Initial design : initial values of the design variables at the beginning of the optimization process,

Objective function(s) : mathematical formulas of a design objective or, in accordance with the given requirements, of several design objectives,

Optimization algorithm : mathematical methods for constrained/unconstrained optimization (optimality criteria methods, mathematical programming methods, hybrid methods),

Optimization model : a comprehensive term for design and evaluation model,

Optimization procedure : a complete software system for dealing with an optimization problem,

Optimization strategy : approach used for reducing complex optimization problems to simplified substitute problems or smaller subproblems,

Preference function : transformation of multiple objective functions into one scalar substitute objective function,

Sensitivity analysis : calculation of the derivatives of the objective function(s) and the constraints with respect to the design variables,

State variables : response of the structural model,

Structural model : mathematical description of the structural behaviour (mathematical-physical model),

Structural parameters : parameters of the structural model,

Multicriteria problem : optimization problem with multiple objective functions.

2.2 Mathematical definitions

In order to solve optimization problems by means of structural optimization methods it must be possible to quantify the design objectives as well as the side-conditions or constraints which have to be expressed by mathematical functions. One example of a design objective is the demand for the highest possible stiffness of a structure. This demand can be described mathematically by the objective "minimization of the maximum deformation of a component". The design variables are the quantities of the structure which have to be varied, e.g. cross-sectional and geometric quantities, the values of which have to be chosen in such a way that the objective function is minimized taking the constraints into consideration. The constraints are the equations of inequalities which contain the mathematical formulation of requirements like design stress, stability criteria, etc. which are not considered in the objective function. The general formulation of the design problem reads

> The values of the design variables have to be chosen in such a way that an objective function $f=f(\mathbf{x})$ leads to an extreme value while taking the constraints into consideration.

This can be expressed mathematically as follows:

$$\min_{\mathbf{x} \in X} \left\{ f(\mathbf{x}) \in \mathbb{R} \right\} \tag{2.1}$$

with feasible domain or the design space:

$$X := \left\{ \mathbf{x} \in \mathbb{R}^n \mid \mathbf{x}_l \leq \mathbf{x} \leq \mathbf{x}_u;\ \mathbf{h}(\mathbf{x}) = \mathbf{0};\ \mathbf{g}(\mathbf{x}) \leq \mathbf{0} \right\}. \tag{2.2}$$

and
- $\mathbb{R}^n$ n-dimensional EUCLIDean space,
- f objective function,
- $\mathbf{x}$ vector of n design variables,
- $\mathbf{g}$ vector of p inequality constraints,
- $\mathbf{h}$ vector of q equality constraints,
- $\mathbf{x}_l, \mathbf{x}_u$ lower and upper bounds of design variables.

In the optimization of mechanical components, the objective function and the constraints are in most case nonlinear functions of design variables. This means, that to find an optimal solution can be very complex, and it is problem-dependent (Nonlinear Scalar Optimization Problem NSOP). If several objectives shall be minimized simultaneously, a Vector or Multicriteria Optimization Problem (MCOP) exists [5]. An example for this is the simultaneous demand for maximum stiffness and minimal structural weight. In this case, the design variable vector has to be determined in such a way that all components of the vector of the objective function become as small as possible and, at the same time, the constraints are met. That means mathematically:

$$"\min_{\mathbf{x} \in X}" \left\{ \mathbf{f}(\mathbf{x}) \right\} \tag{2.3}$$

with $\quad \mathbf{f}(\mathbf{x}) = \left\{ f_1(\mathbf{x}), f_2(\mathbf{x}), ..., f_k(\mathbf{x}) \right\} \quad$ vector of k objective functions.

It is characteristic for multicriteria optimization problems that conflicting objectives occur, i.e. no design variable vector allows a simultaneous optimal fulfillment of all objectives (therefore the expression "min" is used in equation (2.3)). A MCOP is solved by means of transforming the initial problem to scalar substitute problems by so-called preference functions. These substitute problems yield compromise solutions from which the engineer can choose a solution he prefers. A detailed description of the solution strategies for a multicriteria optimization problem can be found in [5, 20].

2.3 Software system requirements

The solution of optimization problems requires software systems which are easy to use, provide sufficient efficiency, and are available for practical applications. Several optimization algorithms should be linked to structural analysis procedures in a suitable manner by means of optimization model processors.

In general, a software system should meet the following requirements:

- possibility of selecting the suitable optimization algorithm for an optimization problem from a number of efficient methods;
- use of different methods for structural analysis such as discretized and analytical methods;
- application of automatic design and evaluation models, i.e. pre- and postprocessors for a wide range of standard problems in optimization modeling, simple integration of special optimization models if required;
- modularized architecture with standard interfaces between the different program modules;
- extensions of the program system by integrating additional modules without comprehensive implementation work;
- reduction of the numerical effort particularly for large structural optimization tasks by applying efficient algorithms (e.g. sensitivity analysis of FE-structures, methods for solving systems of linear and nonlinear equations etc.);
- efficient data management for large optimization problems;
- utilization of modern programming techniques (parallel computing);
- supporting facilities for input data generation and output documentation.

3. STRUCTURE OF THE OPTIMIZATION PROCEDURE SAPOP

3.1 Three-Columns-Concept

In order to deal with optimization problems in the structural design process, a procedure following the "Three-Columns-Concept" seems most suitable. Following this concept, it is most sensible to divide the problem into the "three columns" (Fig. 3/1) [5]:

I   Structural model,
II Optimization model and
III Optimization algorithm.

SAPOP: optimization procedure for multicriteria structural design

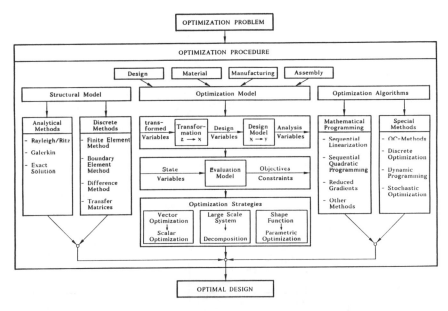

Fig. 3/1. Three-Columns Concept for solving optimization problems

The first step is the theoretical formulation of the optimization problem taking into account all relevant demands on the structure. The next step involves the solution of the subproblems "structural modeling" and "optimization modeling". From the third column an optimization algorithm is selected and linked with the structural and the optimization model in order to form an optimization procedure. In the following a detailed description of the columns is given.

Column I: Structural model

Any structural optimization requires the mathematical determination of the physical behaviour of the structure. In the case of mechanical systems, this refers to the typical structural reponse subject to static and dynamic loadings such as deformations, stresses, eigenvalues, etc.. Furthermore, information on the stability behaviour (buckling loads) has to be determined. All state variables required for the objective function and constraints have to be provided. The structural calculation is carried out using efficient analysis procedures such as the finite element method or transfer matrices methods. In order to ensure a wide field of application, it should be possible to employ several structural analysis methods. Table 3-1 presents some structural analysis programs currently available in SAPOP.

Column II: Optimization model

From an engineer's point of view, this column is the most important in the optimization procedure. First of all, the analysis variables which are to be changed during the optimization process are selected from the structural parameters. The

Name/Literature	Mathematical method	Application
ABSOR [8]	Transfer matrices	Branched shell of revolution; linear statics with non-axially symmetric loads and orthotropic material behaviour
ANSYS [2]	Finite element method	nonlinear statics, dynamics, thermal analysis, etc.
DYNOST [9]	Transfer matrices; Boundary element method	branched shell of revolution; linear dynamics, fluid structure interaction
FEDAS, FEMADA [23]	Finite element method	beam, plate, cylinder; linear statics and dynamics with visco-elastic material behaviour
FIZYL [5]	Analytical method	thick-walled cylinder; linear thermal analysis
LSAP [11]	Finite difference method	shell; linear statics with anisotropic material behaviour
NLPLAT [17]	Finite difference method	rectangular plate; nonlinear statics with anisotropic visco-elastic material behaviour
SAP V-2 [19]	Finite element method	static and dynamic analysis
WABI [7]	Analytical method	cylindrical shell; linear statics with non-axially symmetric load and anisotropic material behaviour

Table 3-1. Structural Analysis Programs available in SAPOP

design model including variable linking, variable fixing, shape functions etc. provides a mathematical link between the analysis variables and the design variables. In order to increase the efficiency and improve the convergence of the optimization, the optimization problem is adapted to meet the special requirements of the optimization algorithm by transforming the design variables into transformation variables, e.g. $\mathbf{z} = 1/\mathbf{x}$. By using this approach, it is possible to almost linearize the stress constraints of a sizing optimization problem. Additionally, objective functions and constraints have to be determined by procedures which evaluate the structural response or state variables. When formulating the optimization model, the engineer has to consider the demands from the fields of design, material, manufacturing, assembly and operation. More details can be found in 3.3.

Column III: Optimization algorithms

In SAPOP, mathematical programming algorithms have been preferred to other methods for solving nonlinear constrained optimization problems. These algorithms are iterative procedures which, proceeding from an initial design $\mathbf{x}_0$, generally

SAPOP: optimization procedure for multicriteria structural design 213

Name	Description		Classification
Search (EXTREM) [14]	Transfomation into an unconstrained problem by means of a penalty function. Problem solution via series of primary and secondary search directions (GRAM-SCHMIT orthogonalization procedure). Optimal step length in the line-search by quadratic interpolation.	- order - characteristics - application	: order 0 : a high number of function evaluations : small and linear/nonlinear
SLP (SEQLI) [5, 15]	Sequential linearization of the nonlinear problem. Solution of the linear subproblem by means of SIMPLEX-procedure. Discrete design quantities are considered using a limited enumeration in the linear subproblem. Solution point is initial point of the proceeding linearization of the nonlinear problem.	- order - characteristics - application	: first order : high efficiency, low effort of formulation : small and large problems, discrete-continuous problems.
SQP (VMCWD) [18]	Sequential linearization and quadratic approximation of the nonlinear problem by means of the BFGS-formula of the HESSE matrix of the LAGRANGE function. Solving of the quadratic subproblem in order to generate the search direction. Optimal step length in the line-search via penalty function and quadratic interpolation.	- order - characteristics - application	: second order : good convergence behaviour, precise optimal point, high effort of formulation : small and average problems.
GRG (GREGA) [1, 3]	One part of the design variables is eliminated from the objective function by means of active constraints. The reduced gradient of the objective function is calculated in order to generate the search direction. Optimal step length in the line-search using a quasi NEWTON algorithm where the reliability is guaranteed in each step.	- order - characteristics - application	: first order : feasible intermediate design, high number function evaluatiors : small and medium problems, highly nonlinear problems.
SQP-GRG (QPRLT) [3, 16]	Generating the search direction by means of the SQP-procedure. Optimal step length by means of the GRG-procedure. The advantages of the SQP- and GRG-procedure are combined in one algorithm.	- order - characteristics - application	: second order : feasible intermediate design, good convergence behaviour, precise optimal point, high storage capacity required : small and large problems, highly nonlinear problems.
Dual method (MMA) [22]	Sequential, convex approximation of the nonlinear problem by means of which the primal problem can be transformed into a dual problem with specific characteristics (separable and convex). The dual subproblem can be solved with the aid of a conventional gradient procedure or a NEWTON algorithm. The solution point is the initial point of the proceeding convex approximation	- order - characteristics - application	: first order : highly efficient, low effort of formulation : small and large problems, highly nonlinear problems, discrete-continuous problems.

Table 3-2. Optimization algorithms available in SAPOP

provide an improved design variable vector $\mathbf{x}_k$ as a result of each iteration k. The optimization is terminated if a breaking-off criterion responds during an iteration. Numerous studies have demonstrated that the selection of the optimization algorithm is problem-dependent. This is particularly important for a reliable optimization and a high level of efficiency (computing time, rate of convergence). If, for example, all iteration results have to lie within the feasible domain, an algorithm that iterates within the feasible domain (e.g. Generalized Reduced Gradients (GRG)) should be applied. Among others, the algorithms shown in Table 3-2 can be applied in SAPOP.

3.2 Optimization loop

Fig. 3/2 shows the interaction of the "three columns" in an optimization loop. First, the decision maker has to describe the structural and optimization model for the special design problem. Based upon an initial design $\mathbf{y}_0$ for the structural variables, the corresponding initial values $\mathbf{x}_0$ for the design variables are determined. The design model then yields the variable subset of the structural parameters to be optimized. These together with the constant structural parameters (material parameters, structural parameters) are taken to define a special design for which the state variables are calculated by the structural analysis. By means of the evaluation model the objective function and constraint values are calculated as one part of the input values for the optimization algorithm. If a special optimization strategy is applied, for example a strategy for solving a multicriteria optimization problem, the behaviour functions and their derivatives are transformed into corresponding substitute values [5]. Otherwise, they are directly transferred to the optimization algorithm. Using this information, the optimization algorithm calculates

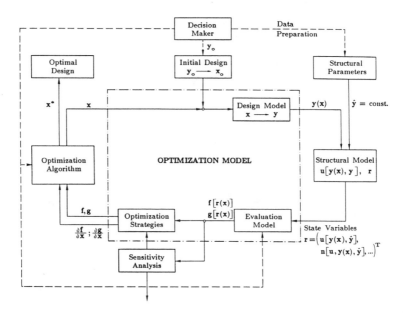

Fig. 3/2. Structure of optimization loop

a new design variable vector and, thereby, one obtains a closed optimization loop. If the optimal design is achieved, which is indicated in the optimization algorithm by breaking-off criteria, the optimization loop is terminated.

In addition to the functional values, most optimization algorithms require gradients of the behaviour functions for the determination of search directions, and approximation models with respect to the design variables which are evaluated by the sensitivity analysis. In addition, gradients are needed for other purposes such as:

- getting more insight into the structural behaviour (system identification),
- minimizing the size of a design model by choosing natural design variables,
- sensitivity of the optimal design relating to non-optimized parameters and
- application of decomposition strategies.

Introducing the state variables $\mathbf{r} \in \mathbb{R}^n$ the objective functions and constraints read as follows:

$$\begin{aligned} \mathbf{f} &= \mathbf{f}(\mathbf{r}, \mathbf{x}), \\ \mathbf{g} &= \mathbf{g}(\mathbf{r}, \mathbf{x}) \end{aligned} \qquad \text{with} \quad \mathbf{r} = \mathbf{r}(\mathbf{x}) = \mathbf{r}[\mathbf{y}(\mathbf{x}), \hat{\mathbf{y}}] . \tag{3.1}$$

The total differentiation yields:

$$d\mathbf{f}(\mathbf{r}, \mathbf{x}) = \frac{\partial \mathbf{f}}{\partial \mathbf{r}} d\mathbf{r} + \frac{\partial \mathbf{f}}{\partial \mathbf{x}} d\mathbf{x} \tag{3.2}$$

with $\frac{\partial \mathbf{f}}{\partial \mathbf{r}} \equiv \left[\frac{\partial f_i}{\partial r_j} \right]_{m \times n}$.

By eliminating $d\mathbf{r}$ it is possible to determine the sensitivity matrix $\mathbf{A}_f$ which corresponds to the derivatives $\partial \mathbf{f}/\partial \mathbf{x}$:

$$d\mathbf{f} = \mathbf{A}_f d\mathbf{x} . \tag{3.3}$$

The derivatives for the sensitivity analysis are usually calculated by the following procedures [13]:

- numerical methods by means of finite differences,
- analytical methods,
- semi-analytical methods (especially for structural analysis by the Method of Finite Elements) and
- hybrid methods.

3.3 Details of the optimization models

3.3.1 Design model

During the development of the SAPOP program system, a fundamental aim was to implement pre- and postprocessors for standard optimization modeling problems. This section gives a brief description of the design and evaluation models. The task of a design model is to calculate the analysis variables $\mathbf{y}$ from the design variables $\mathbf{x}$ by using a unique mapping rule:

$$\mathbf{y} = \mathbf{f}(\mathbf{x}) , \quad \mathbf{x} \in \mathbb{R}^n, \ \mathbf{y} \in \mathbb{R}^{n_y} . \tag{3.4}$$

The analysis variables are a subset of the structural parameters which are required to describe the physical behaviour. In structural optimization problems, the analysis variables are usually sizing quantities (e.g. thicknesses, cross-sections, moments of inertia), geometrical dimensions (e.g. length, width, height), shape parameters (e.g. control points of shape funtions) or material quantities (e.g. E-modul). Particularly when using discrete structural analysis methods (e.g. FE-methods, methods of finite differences), it is important that all structural elements as well as the node topology are determined by the design model. We distinguish the following kinds of models:

a) Linear design model

One part of design modeling can be carried out by linear mapping

$$\mathbf{y}(\mathbf{x}) = \mathbf{A}\mathbf{x} + \mathbf{y}_o . \tag{3.5}$$

Matrix $\mathbf{A}$ is the coordination matrix; the vector $\mathbf{y} \in \mathbb{R}^{n_y}$ is a constant vector. Due to a special structure of the coordination matrix, various design models can be realized. Variable linking is achieved if only one element in each row has the value "one" and if all other elements are zero. The columns of the matrix, however, contain several "one"-elements. Each analysis variable corresponds to only one design variable whereas a design variable can be allocated to several analysis variables. The addition of a constant element to a design variable allows one to consider a constant part of an analysis variable. With variable fixing, a part of the analysis variable vector equals just this constant element, i.e. all elements of the corresponding rows of the coordination matrix are zero. Another possibility is the superposition of design variables to an analysis variable where several elements of a row obtain the value "one". The more general case of this design model is the linear transformation with the analysis variables as linear functions of the design variables [6]. In SAPOP a design model with variable linking and variable fixing is employed for all design variables.

b) Geometric design model

As far as a geometrical optimization is concerned, the design variables represent the components of the vector between two given points of the structure. By adding these components and the respective constant elements, all points of a structure can be defined. This corresponds to the design model of the superposition. It is also possible to interpolate intermediate points. The coefficients of the coordination matrix are determined by means of simple geometrical considerations. For shape optimization problems, SAPOP provides various shape functions to define the surface shape of a component. They allow the transformation into a parametric optimization problem [3, 6, 17]. Among others, the following approach functions are used: LAGRANGE polynomials, BEZIER curves and surfaces, B-spline curves and surfaces, modified ellipse functions etc.. In order to determine the free coefficients of most of the shape functions, control points are employed. The coordinates of these control points are determined from the design variables by using the design model for geometrical optimization described above.

3.3.2 Coupling of design model and structural model

One aim of integrating commercial pre- and post-processors into optimization models therefore is the reduction of the coupling between the structural and the design model. Here one tries to achieve a concept which presents the design model as a mere geometry model. These models can be compared to those of CAD-systems, a fact which facilitates the formulation of the design model largely without the necessity of information about the structural model, e.g. the discretization of the structure. In this model, the body is described as being combined of simple geometrical elements like design points, design lines, design areas and design volumes (Fig. 3/3). The basis for this is the mathematical formulation of these elements in the form of parametrical functions which are able to describe arbitrary contours. By using these design variable definitions, the model yields a (geometrical) design model with the following characteristics:

- arbitrarily precise approximation of the geometry by using parametrical functions,
- description of the geometry by as few degrees of freedom as possible, combined with a low effort of input,
- direct geometrical interpretation of those design variables defined by geometry points,
- simple geometrical assignment of those design variables defined by so-called "attributes",
- graphical-interactive support and control of the modeling as all design quantities are assigned geometrically.

The generation of a useful structural model (FE model) on the basis of the described geometrical model requires, in a further step, the subdivision of the body into finite elements by automatic working mesh-generators [2].

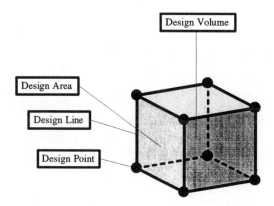

Fig. 3/3. Design Model

3.3.3 Evaluation models

The expression "state variables" **u** of a structural mechanical system refers to quantities such as deformations, stresses, resulting forces, strains, eigenvalues etc.. These state variables depend on the design variables through the structural model. The task of the evaluation model is to formulate the objective function vector **f** and the constraint vector **g** as functions of these state variables. Generally, the constraints represent limiting demands on the state variables; i.e. the quantities of a state variable u_i should be bounded by upper and/or lower limits. This demand provides two constraints which are normalized for numerical reasons:

$$u_i^l \le u_i \le u_i^u \quad \longrightarrow \quad g_1 = \frac{u_i}{u_i^u} - 1 \le 0, \quad g_2 = 1 - \frac{u_i}{u_i^l} \le 0 . \tag{3.6}$$

The limiting values u_i^l and u_i^u can also depend on the design variables. SAPOP calculates the deformation, eigenvalue, stress and buckling constraints in order to optimize structures. As far as composite optimization is concerned, various failure criteria are available (fiber break, max-stress or max-strain criterion, failure criteria by Tsai-Wu etc.). For large structures it is advantageous to reduce the number of constraints. For this purpose, SAPOP provides a "constraint linking strategy" which determines the most critical constraint value of a certain constraint type in a defined domain. Further constraints are upper and lower bounds of the design variables limited by construction, manufacturing etc..

The majority of papers published on examples of structural optimization deal with the minimization of the structural weight. However, for many applications other objectives can also be important like the minimization of deformations of highly accurate systems, the minimization of stresses in order to reduce stress concentrations, or the maximization of certain eigenvalues of a structure. For this reason, SAPOP provides different types of objective functions, i.e. all state variables necessary for the formulation of constraints can also be used as an objective. Due to the nonlinearity of these objective functions, they are more difficult to treat then structural weight.

4. DESCRIPTION OF THE MAIN MODULES

The software system SAPOP consists of three independent columns mentioned previously which communicate via a data-management-system. Every column is divided into single subsystems. Standardized interface definitions exist and modularity of the program system is guaranteed. All subsystems contain a certain number of program modules which are interchangeable. For an optimization calculation, only those modules are addressed which are actually needed.

4.1 Input module

The input system is used for preparing the input data provided stored on a database by the user. All quantities required to describe the structural and the optimization model as well as the parameters for controlling the optimization process are edited here. The user has to provide at least two different data items. The data file OPTDAT includes all data necessary to control the optimization process as well as

SAPOP: optimization procedure for multicriteria structural design

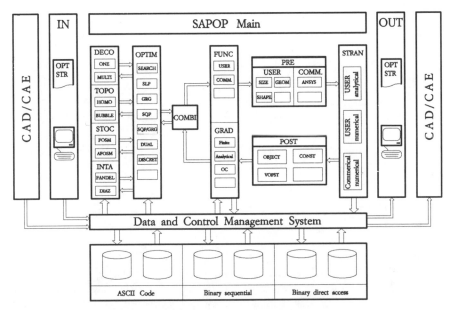

Fig. 4/1. Block diagram of the program system SAPOP

the initial values of the design variables and the input quantities for the formulation of the optimization model. The data file STRDAT includes the input data for the software system which is applied to the structural analysis. In the future, the user will be supported by an CAD/CAE system when generating the input data.

4.2 Main module

The actual optimization computation is carried out by the SAPOP main module MAIN. First of all, an initialization phase is run, and, subsequently, the optimization is started via the one-level-block (ONE) or multi-level-block (MULTI) in the DECO module. Furthermore, various control modules for topology (TOPO), stochastic (STOC) and interactive-vector-optimization (INTA) have been implemented. They can be applied depending on the optimization problem. A number of different optimization algorithms can be called upon by all of these control modules. Additionally, it is possible to couple different algorithms by means of a series connection (serial hybrid approach).

For each iteration the actual values of the objective function and constraints are required, and for most of the algorithms the gradients have to be calculated with regard to the transformation variables. The control program FUNC for structural analyses and the control programs for sensitivity analyses GRAD are called upon via the interface module COMBI.

The subsequently called PRE modules contain different design models used for determining the analysis variables from the design variables. The design model SIZE includes variable linking and variable fixing for cross-section optimization.

SHAPE and GEOM modules can be employed for shape and geometry optimization tasks. If a special design model is to be used to solve an optimization problem, a corresponding program module can be included, or the entire pre-processor can be exchanged. In addition to these USER-programming pre-processors it is possible to choose commercial pre- and post-processors. For this purpose, SAPOP currently provides a special interface to ANSYS/PREP and ANSYS/POST. The structural analysis is now carried out using the updated analysis variables. These are parts of the structural parameters of the mathematical-mechanical model which describes the physical behaviour of the actual structure. Systems of algebraic or differential equations are solved by using efficient numerical methods. Apart from these programs, the user can link his own structural analysis programs to SAPOP. Thus, it is possible to deal with structures using analytical calculations or to deal with any examples from interdisciplinary fields. For the latter, however, other pre- and post-processors are usually required in order to formulate the design and evaluation model.

If the range of performance of the post-processor is not sufficient for a special application, user-defined programs for the formulation of objective functions and constraints can be linked via standardized interfaces. The updated objective functions and constraints are transferred to the optimization algorithm via the modules FUNC and COMBI. For a sensitivity analysis the PRE-STRAN-POST-loop is run several times before the gradients of the objective function and constraints are available. The data exchange between the individual blocks of the SAPOP main module is carried out by the Data Management System. Thus, all program segments have fast access to a large dataset which cannot be kept in core completely. The storage capacity is efficiently used and modularity is ensured. The Data Management System contains input and output control routines to read and write direct access files whereby the data items are identified by a character specification.

4.3 Output module

The output module enables the user to interpret the optimization results, such as history of iteration, breaking-off criteria, constraints, design variables etc.. Two output files are available to the user. The file OPTDAT (ASCII-format) contains the iteration course and the file STRDAT (binary format) contains the output data of the structural analysis method. Future developments aim at a linkage of CAD/CAE to FEM which means in this case that the data of the optimization procedure can be transmitted back to a CAD/CAE by defined interfaces.

5. Application examples

After presenting the fundamentals and realization of the optimization procedure SAPOP, now a list of application examples which are optimized in connection with industrial and public research purposes follows. Table 5-1 offers a selection of examples for which the most important information concerning the previously mentioned Three-Columns-Concept is given. In addition, there exist many other examples which are not mentioned here. This collection of examples does not give a complete description of the features of the optimization procedure SAPOP because the selection of optimization models and different optimization algorithms is possible.

SAPOP: optimization procedure for multicriteria structural design

Examples	Structural Model	Optimization Model	Optimization Algorithm
Fiber-reinforced cylindrical shell under nonsymmetrical load with respect to the axis [7]	Solution: analytical Program: WABI DOF: u,v,w; N,Q,M	$f(\mathbf{x})$: k=3 (dead weight, cost, deformation) $g(\mathbf{x})$: p=200 (composite-failure, laminate and layer thicknesses) $\mathbf{x}$: n=11 (layer thicknesses, fiber angles) Optimization Strategy: MCOP solved by COT and OW	SLP (SEQLI) SQP (VMCWD)
Fiber-reinforced spherical shell under constant internal pressure and temperature [11]	Solution: discrete Program: LSAP DOF: u,v,w; ψ_φ, ψ_ϑ; N,Q,M	$f(\mathbf{x})$: k=2 (dead weight, deformation) $g(\mathbf{x})$: p=3 (composite-failure, laminate thickness) $\mathbf{x}$: n=4 (layer thicknesses, fiber angles) Optimization Strategy: MCOP solved by COT and MM	SLP (SEQLI)
Fiber-reinforced plate with time dependent material behaviour; large deformation [17]	Solution: discrete Program: NLPLAT DOF: u,v,w; ψ_x, ψ_y; N,Q,M	$f(\mathbf{x})$: k=2 (dead weight, deformation) $g(\mathbf{x})$: p=3 (composite-failure, laminate thickness) $\mathbf{x}$: n=4 (layer thicknesses, fiber angles) Optimization Strategy: MCOP solved by COT	SLP (SEQLI) SQP (VMCWD)
Fiber-reinforced sandwich plate under dead weight and pressure load [20]	Solution: FEM Program: SAP V-2 Elements: 3D-Solid, Plate	$f(\mathbf{x})$: k=2 (dead weight, deformation) $g(\mathbf{x})$: p=4 (composite-failure, shear-failure, sandwich thickness) $\mathbf{x}$: n=7 (layer thicknesses, fiber angles and core height) Optimization Strategy: MCOP solved by COT and IAO	SLP (SEQLI) SQP (VMCWD) GRG (GREGA)

MCOP: Multicriteria Optimization Problem;   MLO: Multi Level Optimization;   COT: Method of Constraint Oriented Transformation;
SCOP: Scalar Optimization Problem;   STO: Stochastic Optimization;   OW: Method of Objective Weighting;
   IAO: Interactive Optimization;   MM: Method of Min-Max Formulation.

Table 5-1a: Application examples optimized by SAPOP

Examples	Structural Model	Optimization Model	Optimization Algorithm
Thick-walled fiber-reinforced beam under single load [12]	Solution: analytical Program: CBEAM DOF: u,v,w; σ	$f(\mathbf{x})$: k=2 (cost, deformation) $g(\mathbf{x})$: p=6 (composite-failure, laminat thickness, thermal deformation, production) $\mathbf{x}$: n=11 (layer thicknesses, fiber angles) Optimization Strategy: MCOP solved by COT	SLP (SEQLI) SQP (VMCWD)
Ceramic point-supported mirror plate under dead weight, constant pressure and temperature load [10]	Solution: analytical Program: DISC DOF: w; σ	$f(\mathbf{x})$: k=2 (deformation/surface accuracy, dead weight) $g(\mathbf{x})$: p=3 (deformation, failure-criteria) $\mathbf{x}$: n=6 (surface layer thickness, rib thickness, thickness of boundary stiffening, core height, cell size) Optimization Strategy: MCOP solved by COT and STO	GRG (GREGA) SQP (VMCWD) SQP-GRG (QPRLT)
Thin walled sperical shell with damping materials for vibration reduction [22]	Solution: FEM Program: FEMDA Elements: 2D-Solid	$f(\mathbf{x})$: k=1 (maximum responses in a frequency band) $g(\mathbf{x})$: p=1 (volume) $\mathbf{x}$: n=40 (layer thicknesses) Optimization Strategy: SCP	GRG (GREGA) SQP-GRG (QPRLT)
Chilled cast-iron roller under dead weight, external pressure and temperature load [5]	Solution: analytical Program: FIZYL DOF: u(r,Φ); σ	$f(\mathbf{x})$: k=3 (dead weight, failure-criteria, deformation) $g(\mathbf{x})$: p=3 (stress, geometry, deformation) $\mathbf{x}$: n=4 (thickness of the martensite layer, radius of the borehole, bolt radius, inner radius) Optimization Strategy: SCOP and MCOP solved by COT	SLP (SEQLI) GRG (GREGA)

MCOP: Multicriteria Optimization Problem;   MLO: Multi Level Optimization;   COT: Method of Constraint Oriented Transformation;
SCOP: Scalar Optimization Problem;   STO: Stochastic Optimization;   OW: Method of Objective Weighting;
   IAO: Interactive Optimization;   MM: Method of Min-Max Formulation.

Table 5-1b: Application examples optimized by SAPOP

SAPOP: optimization procedure for multicriteria structural design   223

Example	Structural Model	Optimization Model	Optimization Algorithm
Parabolic reflector for a telescope under dead weight, wind load and temperature load [5]	Solution: FEM Program: SAP V-2 Elements: Spar, Beam, Mass	$f(\mathbf{x})$: k=9 (surface accuracy, deformation) $g(\mathbf{x})$: p=9 (surface accuracy, deformation, weight manufacturing tolerances) $\mathbf{x}$: n=7 (cross-sections, height) Optimization Strategy: MCOP solved by OW	SLP (SEQLI) GRG (GREGA)
Pipe flange joint under single load and single moment [4, 15]	Solution: FEM Program: SAP V-2 Elements: 3D-Solid, Plate/Shell	$f(\mathbf{x})$: k=1 (cost) $g(\mathbf{x})$: p=33 (stress, deformation, screw force) $\mathbf{x}$: n=40 (wall thicknesses, flange height, height and diameter) Optimization Strategy: SCOP	SLP (SEQLI)
Belt conveyer cylinder under non-symmetrical load [5, 15]	Solution: FEM Program: SAP V-2 Elements: Plate/Shell Beam	$f(\mathbf{x})$: k=2 (dead weight, stress) $g(\mathbf{x})$: p=3 (stress, wall thickness) $\mathbf{x}$: n=4 (contour, wall thicknesses) Optimization Strategy: MCOP solved by MM	SLP (SEQLI)

MCOP: Multicriteria Optimisation Problem; MLO: Multi Level Optimisation; COT: Method of Constraint Oriented Transformation;
SCOP: Scalar Optimisation Problem; STO: Stochastic Optimisation; OW: Method of Objective Weighting;
  IAO: Interactive Optimisation; MM: Method of Min-Max Formulation.

Table 5-1c: Application examples optimized by SAPOP

Example	Structural Model	Optimization Model	Optimization Algorithm
Curved beam under single load [3]	Solution: FEM Program: SAP V-2 Elements: 2D-Solid	$f(\mathbf{x})$: k=9 (dead weight, stress) $g(\mathbf{x})$: p=9 (deformation, dead weight, stress) $\mathbf{x}$: n=7 (thicknesses, shape parameter) Optimization Strategy: MCOP solved by COT and MLO	GRG (GREGA) SQP-GRG (QPRLT)
Automotive wheel under rolling bench test, rotating bending test and air pressure [8]	Solution: Transfer matrices Program: ABSOR DOF: u,v,w; ψ; N,Q,M	$f(\mathbf{x})$: k=1 (dead weight) $g(\mathbf{x})$: p=150 (stress, deformation) $\mathbf{x}$: n=13 (shell thicknesses) Optimization Strategy: SCOP and MLO	SLP (SEQLI) SQP (VMCWD)
Liquid filled satellite tank with internal pressure [9]	Solution: Transfer matrices + BEM Program: DYNOST DOF: u,v,w; ψ; N,Q,M Elements: Boundary Elements (liquid)	$f(\mathbf{x})$: k=2 (dead weight, frequency) $g(\mathbf{x})$: p=50 (stress, volume, frequency, deformation shape) $\mathbf{x}$: n=20 (wall thicknesses, shape parameters) Optimization Strategy: MCOP solved by COT	SQP (VMCWD)

MCOP: Multicriteria Optimization Problem;   MLO: Multi Level Optimization;   COT: Method of Constraint Oriented Transformation;
SCOP: Scalar Optimization Problem;   STO: Stochastic Optimization;   OW: Method of Objective Weighting;
   IAO: Interactive Optimization;   MM: Method of Min-Max Formulation.

Table 5-1d: Application examples optimized by SAPOP

6. CONCLUSION

This contribution introduces an optimization procedure for solving optimization problems with one or multiple objectives from the field of structural mechanics as well as the resulting software implementation. Starting from the Three-Columns-Concept with the Columns "Structural Model", "Optimization Algorithms" and "Optimization Model" and by considering other demands, the program system SAPOP (Structural Analysis Program and Optimization Procedure) was developed. Apart from an input system for data preparation and a commercial graphic module for the evaluation of optimization results, the main part of this program system comprises several subdomains. Each of these subdomains has program modules which are interchangeable depending on the requirement of the optimization problem. Thus it is possible to link different optimization strategies in order to solve special optimization problems (multicriteria optimization, interactive optimization, multilevel optimization and stochastic optimization).

The efficiency of the SAPOP optimization procedure presented here is proved by numerous successfully solved optimization problems of different types. Up to now, medium optimization problems (up to 100 design variables and 300 constraints) have been solved. However, the size of prospective optimization problems strongly depends on the efficiency of the hardware environment. The present version of the SAPOP program is implemented on different computers (VAX, SUN, HP and PC). On account of its modularity and efficient data management system, SAPOP is optimally used by workstation computers and personal computers.

SAPOP represents an efficient optimization procedure which may be improved and extended for future applications. Further steps at development comprise basic research as well as the extension and maintenance of the software. Established future aims are strongest possible independence from hardware, installation on parallel computers with shared or distributed memory systems, the extension of the coupling with CAD/CAE-systems (standardization of interfaces), and the generation of a user-friendly graphical interface for the interactive, problem-dependent definition of the three columns and the controlling the optimization procedure.

7. REFERENCES

[1] Abadie, J.; Carpentier, J.: Generalization of the Wolfe Reduced Gradient Method to be Case of Nonlinear Constraints. In: Fletcher, R.: Optimization. Academic Press, New York 1969

[2] ANSYS: Engineering Analysis System. USER MANUAL Revision 4.4 A 1989. Swanson Analysis System, Inc. P.O. Box 65. Houston, Pennsylvania (USA)

[3] Bremicker, M.: Dekompositionsstrategie in Anwendung auf Probleme der Gestaltsoptimierung. Dissertation. Universität-GH Siegen. VDI-Fortschrittsbericht, Reihe 1, Nr. 173, VDI-Verlag, Düsseldorf 1989

[4] Eschenauer, H.: Numerical and Experimental Investigation on Structural Optimization of Engineering Designs. DFG-Research Report. Universität-GH Siegen, 1985. Fries Verlag, Bonn 1986

[5] Eschenauer, H.; Koski, J.; Osyszka, A.: Multicriteria Design Optimization. Procedures and Applications. Springer-Verlag, Berlin, Heidelberg 1990

[6] Eschenauer, H.; Post, P.; Bremicker, M.: Einsatz der Optimierungsprozedur SAPOP zur Auslegung von Bauteilkomponenten. Bauingenieur 63 (1988) 515-526

[7] Eschenauer, H.; Bellendir, K.: Optimal Layout of Cylindrical Composite Shells under Non-Symmetric Loading. ZAMM 72 (1992) T553-T556

[8] Eschenauer, H., Weinert, M.: On the Sensitivity and the Optimal Layout of Complex Shell Structures. ZAMM 72 (1992) T556-T559

[9] Eschenauer, H.; Weinert, M.: Gestaltsoptimierung statisch und dynamisch belasteter Flüssigkeitstanks. In: Abschlußbericht zum BMFT-Vorhaben: "DYNOST - Dynamische Berechnung vorgespannter rotationssymmetrischer Flüssigkeitstanks"

[10] Eschenauer, H.; Vietor, Th.: Some Aspects on Structural Optimization of Ceramic Structures. In: Eschenauer, H.; Mattheck, C.; Olhoff, N. (Eds.): Engineering Optimization in Design Processes. Proceedings of the International Conference Karlsruhe. Nuclear Research Center, Germany Sep. 3-4 1990. Springer-Verlag, Berlin, Heidelberg, New York 1991

[11] Fuchs, W.: Strukturanalyse und Optimierung anisotroper Schalen aus Faserverbundwerkstoff. Dissertation. Universität-GH Siegen. Institut für Mechanik und Regelungstechnik 1986

[12] Geilen, J.: Optimierungsmodellbildung und Auslegung dickwandiger, gemischter Composite-Bauweisen mit stark gekrümmten Bereichen. Dissertation. Universität-GH Siegen, Institut für Mechanik und Regelungstechnik. Erscheint 1992

[13] Haug, E.J.; Choi, K.K.; Komkov, V.: Design Sensitivity Analysis of Structural Systems. Academic Press, Orlando 1986

[14] Jacob, H. G.: Rechnergestützte Optimierung statischer und dynamischer Systeme. Springer-Verlag, Berlin, Heidelberg, New York 1982

[15] Kneppe, G.: Direkte Lösungsstrategien zur Gestaltsoptimierung von Flächentragwerken. Dissertation. Universität-GH Siegen. VDI-Fortschrittsberichte, Reihe 1, Nr. 135, VDI-Verlag, Düsseldorf 1986

[16] Parkinson, A.: Wilson, M.: Development of a Hybrid SQP-GRG Algorithm for Constrained Nonlinear Programming. Proceedings of the ASME Design Engineering Technical Conference. Columbus, Ohio, Oct. 5-8 1986

[17] Post, P.: Optimierung von Verbundbauweisen unter Berücksichtigung des zeitabhängigen Materialverhaltens. Dissertation, Universität-GH Siegen. Fortschrittsberichte, Reihe 1, Nr. 172, VDI-Verlag, Düsseldorf 1989

[18] Powell, M.J.D.: VMCWD: A FORTRAN Subroutine for Constrained Optimization. University of Cambridge, Report DANTP 1982/NA4

[19] SAP V-2: A Structural Analysis Program for Static and Dynamik Response of Linear Systems. University of Southern California, Los Angeles 1977

[20] Schäfer, E.: Interaktive Strategien zur Bauteiloptimierung bei mehrfacher Zielsetzung und Diskretheitsforderungen. Dissertation. Universität-GH Siegen. Fortschrittsberichte, Reihe 1, Nr. 197. VDI-Verlag, Düsseldorf 1990

[21] Schittkowski, K.: Nonlinear Programming Codes - Information, Tests, Performance. Lecture Notes in Economics and Mathematical Systems, No. 183. Springer-Verlag Berlin, Heidelberg, New York 1980

[22] Svanberg, K.: The method of moving asymptotes - A new method for structural optimization. International Journal for Numerical Methods in Engineering 24 (1987) 359-373

[23] Wodtke, H.-W.: Optimale Auslegung von Dämpfungsbelägen zur Schwingungsreduzierung. Dissertation. Universität-GH Siegen, Institut für Mechanik und Regelungstechnik. Erscheint 1992

Authors' address:

Prof. Dr.-Ing. Hans A. Eschenauer;
Dr.-Ing. Johannes Geilen;
Dipl.-Ing. Hans Jürgen Wahl

Research Center for Multidisciplinary Analyses
and Applied Structural Optimization (FOMAAS)
University of Siegen
Paul-Bonatz-Straße 9-11
W-5900 Siegen (FR Germany)
hjwahl@imr-sun2.fb5.uni-siegen.de

SHAPE: A STRUCTURAL SHAPE OPTIMIZATION PROGRAM

Doz. Dr. Erdal Atrek

Abstract: The commercial shape optimization program entitled SHAPE is described. Its scope, capabilities, and algorithm are outlined, together with an overview of the underlying theory. Examples of shape optimization are presented for illustration.

1. Introduction

Current methodology in the area of computational structural analysis by the use of the finite element technique is well established. This approach, when properly applied, is generally accepted by the industry as providing a fair assessment of the structural performance for most practical cases of interest. As a result, the finite element method has become a standard tool for structural analysis.

Based on the success of the finite element method in predicting structural behavior, work has been directed towards advancing methodology for "computational design". Computational design, or optimum structural design as it has come to be known, involves directed design changes based on structural theory and mathematical methods of optimization.

The development of practical tools for application in optimum structural design has been relatively slow due to the complexity of the design process as well as due to the nonlinearity, and at times seeming discontinuity, of the underlying mathematical problem. Only in the second half of the last decade have widely available related commercial programs been marketed with varying success and been able to gain some foothold, usually based on the strength of the underlying finite element analysis package. Even then, widespread acceptance of optimum structural design software still needs increased industry awareness of the potential of such software for improved productivity and quality as well as for cost savings.

Shape optimization, or even basic shape determination, of structural design is an important subject within computational design. Broadly classified, the procedures for shape optimization fall into two, or perhaps three, categories. The methods of one category are involved with appropriately varying design variables that describe the boundaries of the structural shape (Figure 1) and adjusting the finite elements at the boundaries accordingly. While this sounds simple enough in principle, practical application is quite difficult. Mainly, expertise is required to relate the finite element mesh to the design parameters describing the boundaries. This may get hopelessly complicated, especially for complex solid shapes and for large design variations. As a result, this approach currently appears to be more suitable for small and local boundary refinements where the basic shape is relatively well established. Examples of this approach can be found in Cristescu and El-Yafi (1986) and Hou et al. (1990) where the variables are the nodal coordinates of the finite element mesh, in Fleury (1986), Kumar et al. (1989), and Imam (1982) where the variables are positions of control nodes on the parametric curves or surfaces used to model the boundaries, and in Botkin (1981) where the variables are amplitudes of functions that relate nodal points along the boundaries.

The methods of the second category recognize that an optimized shape is defined not only through the boundaries of the original shape, but through the entire domain of a design envelope. Consequently, the optimum design is simply a subset of the design envelope, and can be obtained by appropriately removing material from the design envelope (Figure 2). This concept first seems to have been envisioned by Maier (1973). Developed independently and elaborated upon, it also underlies the theory and algorithm

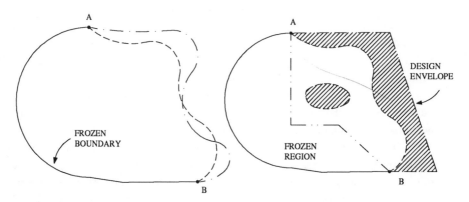

Figure 1. Shape optimization by boundary variation

Figure 2. Shape optimization by material removal

embodied in the SHAPE program (Atrek 1989, 1990, 1991a, 1991b, Atrek and Kodali 1989, Atrek and Agarwal 1992, and Henkel et al. 1991) for the shape optimum design of continuum structures. A rule-based application of this concept may be found in the work of Rodriguez-Velazquez and Seireg (1985, 1988).

A third category related to both the first and the second is based on the so called homogenization method. In application, each finite element can be considered to be a composite cell with an initially undetermined ratio of solid to space (Kikuchi and Suzuki 1991, Bendsoe and Kikuchi 1988) which ratio is then optimized. However, element removal seems to be applied only for condensation rather than during optimization, thus hindering to some degree the advantages that may arise from changes in topology.

Perhaps the greatest advantage to shape optimization by material removal is the ability to rapidly attempt optimization of even the most complex continuum shapes, since related preparation involves little effort beyond that required for a finite element model of the initial shape. Also, very large design changes may be obtained without a need for frequent remodeling. Another important advantage is the ability to create holes and cutouts by removing material from inside the domain of the design envelope. As a special case, material removal can be restricted only to the boundaries at each step, thus providing shape optimization in a manner analogous to the methods of the first category described above. Last, but not least, material removal provides the opportunity to search for the global optimum by the creation of new paths of force transfer.

2. The Program

Scope:

SHAPE is a commercial computer program designed for the shape optimization and sensitivity analysis of complex continuum structures. The program is finite element based both for analysis and shape optimization purposes. Currently it is applicable to the volume, mass, or cost minimization of structures made up of one or more homogeneous materials and exhibiting linear response under multiple static load cases. The types of loading that may be considered include nodal or distributed loads, body forces, and hydrostatic "follower" pressure. Constraints may be placed on nodal displacements and element stresses.

The program will shape optimize continuum structures by removal of material from an initial design envelope. Numerically, the material removal is carried out by the removal of finite elements from the model. Therefore the approach is quite natural to the finite element method. Solids, modeled by solid elements as well as by axisymmetric elements (for axisymmetric loading), shells, and planar systems (both plane stress and plane strain models) can be treated by the program.

SHAPE employs mostly the simplest types of finite elements (triangles and tetrahedra) so that the problem size remains in check even when a very large number of small elements is used. The use of small elements allows the program higher accuracy in shape changes. Models created by use of other elements, such as quads or bricks may be converted into those with triangles through the pre-processor DISPLAY (Anon. 1991a) and tetrahedra through SHAPE companion program DIVELM (Atrek 1991b). Accessory elements such as bars, springs and rigid links can also be used for analysis accuracy.

Variables:

Due to the nature of the shape optimization process within the program, there are no explicit variables to be defined by the user. However, internally the program uses the element volumes as design variables. Continuous solutions involving the element volumes are converted to 0-1 decisions regarding removal or retainment of elements. Naturally, the accuracy of such decisions is proportional to the number of the elements used to model a given structure. On the other hand, experience indicates that even coarse models may yield good approximations to the optimum shape. The shape may then be reworked and remodeled with a finer mesh for further refinement.

Constraints on the Response Quantities:

The program currently allows the user to place constraints on element stresses and/or nodal displacements. The stress constraints may be on the components of the stress tensor, as well as on the principal stresses, maximum shearing stress, or the von Mises equivalent stress. Displacement constraints can be on the translations and rotations (where applicable) at nodal degrees of freedom, as well as on resultant and/or relative displacement quantities. Linear combinations of displacements at nodal degrees of freedom may also be constrained.

The constrained stress and displacement quantities may have different positive and negative limiting values, if and when applicable. For example, the von Mises equivalent stress is always a positive quantity, but principal stresses can have positive or negative values. Similarly, a displacement component may be constrained with different limiting values in the positive and negative directions.

There is no practical limit on the number of constraints to be imposed. Constraints may be imposed with different limiting values at different locations within the model, and with different limiting values for different load cases. Due to the compact form of input, several hundred thousand constraints may be specified within a very few lines of input. The program uses an active set strategy whereby the active set dynamically changes both in terms of size and contents for each iteration. It should be noted that all constraints are evaluated whether they are in the active set or not, and they are placed into the active set after being sorted in order of criticalness. The response quantities for which the constraints have been placed in the active set are then subjected to sensitivity analysis, the results of which are used for various purposes.

Constraints on the Physical Model:

These constraints define action to be taken with respect to the design variables, which are implicitly the element volumes. The constraints on the physical model fall into three categories:
- those used for freezing prescribed regions of the design.
- those used for enforcing symmetry where otherwise none would exist.
- those used to link elements together in groups for more uniform removal of material.

The freezing of elements may be necessary for various reasons, one of which may be satisfaction with parts of the initial shape. The freezing can be done by specifying element, node, or material ID numbers associated with the regions to be frozen, the last capability requiring the least effort.

The next two categories of constraints on the physical model are collectively called "variable linking" within the context of SHAPE. While they are almost similar in final application, they do require different handling for the elements sharing edges or surfaces with any planes of enforced symmetry. In either case, elements can be grouped together in prescribed sets. During optimization, all elements in a given set are treated in the same manner with regard to removal or replacement.

Global Optimum:

The algorithm developed for the optimization process seeks to find the global optimum rather than one of the many possible local optima. This is achieved by the use of two concepts which complement each other. The first concept is perhaps best described by means of a simple Venn diagram as in Figure 3.

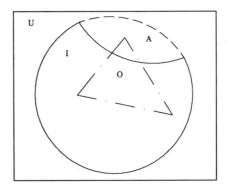

Figure 3. Venn diagram for the concept of the global optimum

Assuming that the design iterate I was obtained by removing material from the initial design envelope U, it is noted that the optimum O may not lie totally within I. This is due to the linearization of the nonlinear problem for solution at design U to obtain I, and also due to the discrete nature of the material removal. Obviously the value of the objective function at I is smaller than that at U, and the shape corresponding to I may still be feasible. However, if O is not contained in I, any further shapes obtained by removing material from I will converge only to a local optimum. Thus, it is clear that an algorithm that seeks the global optimum should expand I to contain O, in this case by adding material.

The second concept therefore is involved with creating the means by which to determine the process of replacing some of the removed material so that the optimum is contained in the augmented set I+A. It is noted now that although the iterate I clearly corresponds to a smaller value of the objective function, there may be a disproportionate change in the values of some of the constrained response quantities while material is removed to obtain I from U. This is again due to the approximate solution employed at the design U. It should also be noted that some constraints which were not in the active set (i.e. which did not seem to be critical) at the design U may now be critical at I. In short, the quality of the shape at I stands improvement if indeed it does not contain the optimum O. Thus what is needed is a measure of the quality of the shape in addition to the objective function. Since this measure has to strike a balance between the objective function and the constraint values, in SHAPE a quantity termed the 'virtual objective function' (such as virtual volume, if the objective function is material volume, or virtual cost if the objective function is the material cost) is defined as

$$W_v = \frac{W}{F_{min}} \tag{1}$$

where W is the actual objective function, and F_{min} is the most critical factor to the constraint surface.

Obviously, if one can truly approach the optimum (the finite element mesh allowing) the value of F_{min} will approach unity, and therefore the virtual objective function will approach W. Based on this, the virtual objective function rather than the actual objective function is used as the true measure of minimization. Looking at Figure 3, it is seen that although the value of W is smaller for I than for U, the value of W_v is expected to be greater, since I does not contain O. Thus, the program can easily conclude that some of the removed material needs to be restored.

The locations where material needs to be restored are found by means of sensitivity analysis, such that the minimum amount of material is restored for maximum effect. Again an active set of the most critical constraints is used for sensitivity analysis. Since it is difficult to determine the actual amount of material to be restored at each location, an iterative process is adopted. This process gives way to the new design step when it is found that it is not possible to decrease the value of the virtual objective function any more. The shape corresponding to the minimum value of the virtual objective function becomes the next envelope U with which optimization will continue.

Sensitivity Analysis:

This facility serves three separate purposes:
1. During the stage ending with material removal, sensitivity analysis is performed for the response quantities in the active set of constraints, in order to set-up the approximate linearized mathematical optimization sub-problem leading to the removal of elements.
2. During the stage characterized by restoral of some of the removed material, sensitivity analysis is performed for the response quantities corresponding to the more critical constraints, in order to find the best locations where elements should be restored for the maximum effect.
3. As an option separate from shape optimization, sensitivity analysis may be performed for a set of user prescribed response quantities. The information is then presented to the user in forms suitable for plotting or browsing. If desired, limiting values may be defined by the user so that the program can sort the response quantities in order of criticalness. Sensitivity analysis results are then found only for a user prescribed or default number of the most critical response quantities.

The sensitivity analysis facility in SHAPE is described in considerable detail by Atrek (1990, 1991b).

Compatibility and Availability:

SHAPE has been designed to be compatible with other commercial finite element analysis software available from Engineering Mechanics Research Corporation. Therefore, the finite element models to be used with SHAPE follow the format for the preparation of input files for the NISA II analysis program (Anon. 1991b). As a result, the relevant input files may be prepared by using the DISPLAY pre-processing program (Anon. 1991a). Similarly, DISPLAY may be used for post-processing the results created by SHAPE. During optimization, for each improvement in shape, the program will create finite element input files in format similar to the original input file.

SHAPE may also be used purely for analysis purposes just as with NISA II, the only difference being that SHAPE will analyze only those models it can shape optimize. Thus, stress and displacement results can be post-processed from SHAPE analysis results. In addition, when employed in the sensitivity analysis mode, SHAPE will enter any required sensitivity analysis results as well into the post-processing files, so that these can be post-processed by DISPLAY in a somewhat similar manner to that for stresses.

The program is available on a commercial basis on most computers in the PC386 to the supercomputer range.

3. Optimization Model

This section describes the theoretical basis underlying the optimization problem.

The Objective Function:

Assuming a continuum structure modeled by finite elements, the total material volume V of the structure is simply the sum of the element volumes v_i:

$$V = \sum_{i=1}^{n} v_i \qquad (2)$$

where n is the total number of finite elements in the model. V becomes the objective function W if the material volume of the structure is to be minimized. In case material mass or cost is to be optimized, the terms of the objective function are simply modified by appropriate factors q_i:

$$W = \sum_{i=1}^{n} q_i v_i \qquad (3)$$

where these factors are determined based on the nature of the objective function.

Constraints:

With the current version of shape, the response quantities on which constraints can be imposed are stresses and displacements. Symbolically, a constraint on a response quantity can be shown as

$$r_j \leq r_j^* \qquad (4)$$

where r_j is the j'th response quantity and r_j^* is the limiting value. When response quantities may take on negative limiting values, the relational symbol in Expression (4) would be reversed and r_j would therefore be bounded from below. The relation given by (4) however is sufficiently general for the rest of the discussion, and will be employed for simplicity.

It can be shown easily that any stress or displacement quantity r_j can be expressed as an exact sum of element contributions c_{ji} at a given design:

$$r_j = \sum_{i=1}^{n} c_{ji} \qquad (5)$$

For example, to express the displacement at a nodal degree of freedom, one may apply a unit virtual force along that degree of freedom and compute the element contributions to that displacement. Thus, each element contribution c_{ji} can be found as the dot product of two vectors

$$c_{ji} = \{x_1\}^T \{x_2\} \qquad (6)$$

where the first vector is the set of element nodal forces due to the actual loading, and the second vector is the set of element nodal displacements due to the virtual loading, or vice versa. The generation of the virtual load vectors, reanalysis for the multiple right hand sides due to the active set of constraints, and the evaluation of the coefficients c_{ji} constitute the sensitivity analysis effort within the optimization stage.

It should be noted that Equation (5) is not used to compute the value of the response quantity r_j, since this is already found through regular analysis, but it is to be used to relate the response quantity to the variables. In this particular case, the variables that are of interest are the element volumes v_i. One can then write, without loss of generality, Equation (5) in terms of the element volumes, in the form

$$r_j = \sum_{i=1}^{n} \frac{b_{ji}}{v_i} \tag{7}$$

where the coefficients b_{ji} can be found by multiplying the contributions c_{ji} by the element volumes v_i. It should be noted that indeed the coefficients b_{ji} are not constants in general, but are assumed to be so only at a given design. Therefore, the formulation does require the recalculation of the coefficients b_{ji} as the shape changes, to avoid divergence during optimization. The formulation succeeds because the element volumes are the major factors in the element contributions, for prescribed loading and boundary conditions. For example, for a truss system the lengths of the bars (in the numerator of the contributions) are constant, and the change in the contributions depend on the change in cross-sectional areas (therefore the volumes), to the first order.

The number of constraints specified for the problem may be way in excess of what is needed at any given stage in the solution. Thus, while all related response quantities are evaluated, optimization proceeds based on an active (or critical) set of constraints.

For frozen elements, the constraint expressions are modified simply by taking the volumes of these elements to be constants rather than variables. Thus, Equation (7) becomes

$$r_j = \underset{\text{frozen}}{\sum c_{ji}} + \underset{\text{removable}}{\sum \frac{b_{ji}}{v_i}} \tag{7a}$$

indicating a decrease in the number of variables to be dealt with.

Variable linking is imposed by grouping variables appropriately in Equation (7), leading to a further reduction in the number of variables.

Optimality Criteria and the Optimization Sub-problem

Substitution from Equation (7) into Expression (4) gives

$$\sum_{i=1}^{n} \frac{b_{ji}}{v_i} \leq r_j^* \tag{8}$$

By associating a Lagrange multiplier λ_j with each such constraint j, the constraint may be converted into an equality:

$$\lambda_j \left(\left(\frac{1}{r_j^*} \sum_{i=1}^{n} \frac{b_{ji}}{v_i} \right) - 1 \right) = 0 \qquad (9)$$

Thus, if the constraint is passive (i.e. the term in parentheses is non-zero) the corresponding Lagrange multiplier has to be zero. Otherwise, if the constraint is active, the Lagrange multiplier is usually non-zero. It may still be zero in case the constraint is only incidental at the optimum.

It is quite unlikely that more than one or two constraints will be active at any design before the optimum is reached (unless, of course, there is symmetry). Indeed, since the shape design cannot be scaled in a similar manner to that for resizing optimization, it is not even possible to make constraints become active at a given design, if the design is still conservative. Thus, it would appear that almost all Lagrange multipliers would end up being zero during solution. This would hamper solution to a great degree, since it would not be possible to employ Equation (9).

The above problem can be avoided if the limiting value is replaced by the actual value of the constrained response quantity. This amounts to making the limiting value variable, and declaring that the most critical constraints are active at a given design. As the design changes, both the set of critical constraints and the actual values change as well, giving great flexibility in approaching towards the optimum. Thus, Equation (9) is replaced at each shape by the following, where now the instantaneous value of the response quantity is used:

$$\lambda_j \left(\left(\frac{1}{r_j} \sum_{i=1}^{n} \frac{b_{ji}}{v_i} \right) - 1 \right) = 0 \qquad (10)$$

Then, if the constraint j appears to be instrumental in the drive towards the optimum, a non-zero Lagrange multiplier will be obtained.

Adding all such constraint expressions to the objective function, one obtains the Lagrangian:

$$L = \sum_{i=1}^{n} q_i v_i + \sum_{j=1}^{m} \lambda_j \left(\left(\frac{1}{r_j} \sum_{i=1}^{n} \frac{b_{ji}}{v_i} \right) - 1 \right) \qquad (11)$$

where m is the number of constraints collected into the active set. It should be noted that writing Equation (11) for all the constraints would be meaningless, because then all the constraints would have been deemed to be active, and there would be no room for progress.

If the shape at which Equation (11) is written were the optimum shape, then one would have

$$\left\{\frac{\partial L}{\partial v_i}\right\} = \{0\} \qquad (12)$$

for the values of the variables v_i and of the response quantities r_j at that particular shape. Here $\{0\}$ is the null vector and Equations (12) are the necessary set of Kuhn-Tucker conditions (Kuhn and Tucker 1951) for optimality.

Computation for Design Change:

Unfortunately, it is not possible to solve Equation (12) directly for a new shape since the number of variables will not match the number of constraints, and even if it did there is no guarantee of a unique solution. Finally, there is no guarantee either that all the constraints in the current active set will be active at the optimum. On the other hand, a somewhat less direct approach is quite rewarding.

It is first noted that at the optimum, the sum of the Lagrange multipliers for the active constraints will be equal to the value of the objective function, i.e.

$$\sum_{j=1}^{m} \lambda_j = W_{\text{optimum}} \qquad (13)$$

Away from the optimum, there is a gap between the sum of the Lagrange multipliers and the objective function, and the gap approaches zero during convergence to the optimum. In other words, the sum of the Lagrange multipliers represents a lower bound to the value of the objective function. Thus, one can form the following optimization sub-problem at the current design:

$$\text{Maximize} \sum_{j=1}^{m} \lambda_j \qquad (14a)$$

$$\text{Subject to} \left\{\frac{\partial L}{\partial v_i}\right\} \geq \{0\} \qquad (14b)$$

where the upper bound for m is the total number of constraints. To gain further insight, the differentiation indicated in Expression (14b) is carried out, giving, upon some rearrangement

$$\sum_{j=1}^{m} \left(\frac{b_{ji}}{r_j}\right) \lambda_j \leq q_i v_i^2 \;, \quad i = 1, \ldots, n \tag{15}$$

which is obviously linear in the Lagrange multipliers.

The problem outlined above is solvable as a linear programming problem in terms of the Lagrange multipliers if the current values of the element volumes v_i are substituted into Expressions (15). Usually, the number of the constraints in the active set is much less than the number of elements. Therefore, it is much easier to solve the dual of the above problem in practice. Once the Lagrange multipliers have been computed, the element volumes corresponding to the "changed" design can be found by converting Expressions (15) into a set of equalities:

$$\bar{v}_i^2 = \frac{1}{q_i} \sum_{j=1}^{m} \left(\frac{b_{ji}}{r_j}\right) \lambda_j \;, \quad i = 1, \ldots, n \tag{16}$$

where the overbar indicates the new values for the element volumes.

This approach is particularly useful in qualifying further those constraints that have a higher chance of becoming active at the optimum. Thus, if the problem is written for a set of critical constraints (say the most critical m constraints), the solution for the problem of Expressions (14) may have zero Lagrange multipliers associated with some of these constraints. This is an indication that the related constraints are currently not really critical in terms of driving the optimization process.

The Basic Algorithm:

The results obtained by the solution of the optimization sub-problem of Expressions (14a) through (16) are applied by converting the solution for the new element volumes into decisions regarding the removal or retainment of each element. For example, due to the squares of the volumes in Equation (16), some element volumes may well come out to be imaginary (i.e. negative values for the squares). These make up the first set of elements to be considered for removal. Additional elements may end being removed as well, based on a computed target value of the objective function at each stage.

Element removal is tempered by considerations related to the integrity of the finite element model. Thus, elements earmarked for removal are retained if their removal would compromise the continuity of the finite element model. As a result, element removal is carried out in very carefully controlled stages, the end result of which is an analyzable finite element model.

Although a shape has now been obtained via the forced satisfaction of the optimality criteria, two items need to be considered if the procedure is to lead to an optimum design. The first of these concerns the nature of the active set. The active set initially is made up of the most critical constraints, with the assumption that the most critical constraints may be instrumental in finding an optimum. However, the Lagrange multipliers may turn out to be zero for some of these constraints, and the analysis of the new shape may indicate new constraints that become quite critical. Thus, it is necessary to update the active set with additional constraints, and to re-solve the optimization subproblem of Expressions (14). Usually, mainly based on the values of the "virtual objective function", the program can select a shape within three or four such re-solutions. The solution of the optimization sub-problem for various updates of the active set, and the selection of one of the generated shapes constitutes the first stage of an optimization design step.

The second item to be considered involves the error associated with the linearization of the actual mathematical problem at the beginning of the design step. Due to the change in shape as a result of the solution for the optimality criteria, the linearization is no longer valid, and the advance has to be evaluated at the generated design. This leads to the replacement of some of the elements that have been removed in the first stage of the design step. The philosophy underlying this second stage of the design step has already been discussed under the sub-heading "global optimum". The element replacement is carried out in an iterative manner based on results of sensitivity analysis for a set of critical constraints at each iteration. The element contribution values of Equation (5) are summed up at the surface (or border) nodes where elements had been removed, for an indication of where material should be replaced for the maximum beneficial effect on the response (e.g. lower stress or displacement). While this is only an approximate evaluation of the sensitivity analysis results, it has proven to be remarkably powerful in rapidly increasing the quality of the design under most circumstances. The value of the "virtual objective function", which is a trade-off between the objective function and response, is tracked during the iterations, and the design step is completed when the lowest possible such value is deemed to have been found. The next design step starts with this shape corresponding to the minimum value of the "virtual objective function".

The macro-flowchart of the algorithm is given in Figure 4.

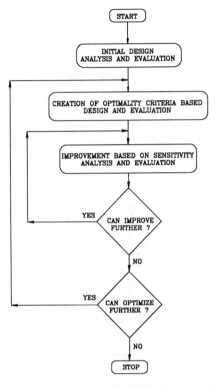

Figure 4. Macro flowchart of the shape optimization algorithm

4. Structural Analysis

The structural analyses required during shape optimization are carried out through the appropriate subroutines of NISA II, incorporated into SHAPE. These routines are used both for the analysis of generated designs as well as in sensitivity analysis with simultaneous solution for multiple right hand sides.

The solution of the linearly static equilibrium equations follows the wavefront solution variation (Irons 1970) of Gaussian elimination. Wavefront minimization is available to increase solution speed, or, at times, to be able to fit a large problem for solution. For solids, the wavefront minimization is best done at the time the brick (hexahedron and wedge) model is being converted into a tetrahedron model with the companion program DIVELM.

As optimization will usually involve multiple load cases, these are solved for in exactly the same manner as in NISA II, except that the boundary conditions are not allowed to change from one load case to the other. This is not due to any difficulty in static analysis, but rather due to the inefficiency related to switching boundary conditions during sensitivity analysis. The difficulty can be circumvented, in such cases, by replacing boundary conditions with equivalent forces, and updating the forces during optimization.

5. Examples

Several examples are given here to illustrate the application of SHAPE to both planar and solid problems. Other examples of the application of SHAPE can be found in the literature (Atrek 1989, 1990, 1991a, 1991b, Atrek and Kodali 1989, Atrek and Agarwal 1992, and Henkel et al. 1991). The examples here range from a primitive two-dimensional approximation of a bridge under a single load case, to the shape optimization of an automobile component under multiple static load cases.

Bridge:

Figure 5 shows the design envelope for a hypothetical simplistic bridge modeled by triangular plane-stress elements. The only load case consists of a uniformly distributed load (q = 10.0) applied under the roadway, representing the self-weight of the roadway. The dimensions of the design envelope are 20 x 10 x 0.1. The roadway and the abutments are frozen, and the rest of the model is free to change during shape optimization. The material constants are Young's modulus = 1.0×10^7 and Poisson's ratio = 0.3.

Figure 5. Design envelope for bridge

Figure 6. Optimum shape of bridge for stress constraints

a) The bridge is shape optimized for an allowable von Mises stress of 50.0 units. The resulting configuration is shown in Figure 6. For this simple problem no distinction has been made between compressive and tensile stresses. However, in the practical design environment, stability considerations require the allowable compressive stress to be related to the slenderness ratio. This can be implemented easily by imposing separate stress limits on the negative and positive values of the principal stresses. The limit on the compressive stress can be modified as the design progresses, based on stability considerations.

b) The same design envelope is shape optimized again, but this time for a limit on the maximum deflection. This limit is chosen as the maximum value obtained for the optimum design of part (a). The resulting configuration is shown in Figure 7. As expected, the stresses for this design are much higher than those for part (a).

Figure 7. Optimum shape of bridge for maximum deflection constraint

c) Optimization for both the stress and deflection constraints resulted in a shape very similar to that of part (a), where the deflection constraint was only incidentally "active" at the optimum and the shape determined by the stress constraints.

Torque arm:

The basic dimensions and the loads for this problem (Figure 8) are the same as those for the torque arm originally shape optimized by Bennett and Botkin (1983). The only difference in the initial shape is that the initial shape of the torque arm in the original reference had an oblong cutout, which has been omitted in the present case in order to investigate what sort of material removal would take place from inside the design envelope. While the original problem had two load cases so that symmetry could be enforced, with SHAPE symmetry was enforced through variable linking so that it was possible to apply only one of the two load cases instead of both. The optimum shape obtained is shown in Figure 9. It is noted that this is quite different than found in Bennett and Botkin (1983), indicating the latitude allowed through removal of material. For this problem, only the material around the two holes was frozen.

Figure 8. Design envelope for torque arm   Figure 9. Final shape obtained for torque arm

Lower control arm:

The optimization of this automobile suspension control arm (Figure 10), allowing shape changes everywhere except for the frozen regions, was reported by Atrek (1989). Here, the same control arm is optimized by removal of material only on the immediate boundaries at each stage. The brick elements of the initial model were automatically converted into tetrahedra for optimization.

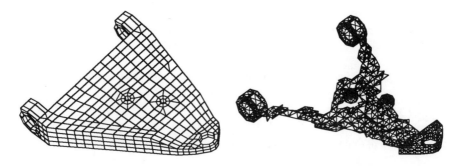

Figure 10. Design envelope for lower control arm

Figure 11. Final shape for boundary optimized lower control arm model

The shape optimization is performed for five separate load cases, with different values for the limiting von Mises stress at each load case. The material at the bushing and the ball-joint connection locations, as well as at the stabilizer bar mounts (roughly at the middle of the control arm) is frozen for obvious reasons. The final shape that could be obtained with the given mesh is shown in Figure 11. While much more improvement is still possible, the solution slows down considerably at this point, indicating the need for a mesh refinement, preferably after a re-modeling of the control arm.

6. Conclusion

The program SHAPE has been designed and developed with the objective of providing the engineer with a practical, straightforward, and powerful tool for the shape determination and optimization of two- and three-dimensional structures. It can be used to determine the rough outlines of the optimum shape as well as to converge towards an accurate representation of the shape with successive mesh refinements. By virtue of its design, it is extremely easy to use, even for very complex shapes. The algorithm used in the program is well tested and quite mature, and is based on solid theoretical foundation.

Since design is a much more complex procedure than analysis, and since large scale numerical shape optimization is a relatively new development, obviously many things remain to be done. Among these one can cite the smoothing of boundaries into manufacturable surfaces, for example by a pattern recognition program, and optimization for various other loading and response regimes, and these are left to the foreseeable future. It also appears that more powerful variations of the present approach can be developed, increasing the practical value of the method. Again one must look into the future for these.

In conclusion, it is believed that SHAPE is a sophisticated, reliable, and robust tool for the engineer who desires faster design cycles and significant weight savings without a sacrifice in structural safety.

References

Anon. (1991a), DISPLAY User's Manual, Engineering Mechanics Research Corporation, Troy, Michigan.

Anon. (1991b), NISA II User's Manual, Engineering Mechanics Research Corporation, Troy, Michigan.

Atrek, E. (1989), "SHAPE: A Program for Shape Optimization of Continuum Structures", Computer Aided Optimum Design of Structures: Applications, (C.A. Brebbia, S. Hernandez, Editors), (Proceedings of the first International Conference, OPTI '89, Southampton, U.K., June 1989), Computational Mechanics Publications, Southampton, Springer Verlag, Berlin, pp. 135-144.

Atrek, E. (1990), "Sensitivity Analysis with the SHAPE Optimization Program", Advances in Design Automation -1990-, Vol. Two: Optimal Design and Mechanical Systems Analysis,

(B. Ravani, Editor), (DE-Vol. 23-2), (Proceedings of the 16th Design Automation Conference, Chicago, Illinois, September 16-19, 1990), ASME, New York, pp. 253-259.

Atrek, E. (1991a), "Structural Design Applications with the SHAPE Optimization Program", NISA Users Conference, Tokyo, Japan.

Atrek, E. (1991b), SHAPE User's Manual, Engineering Mechanics Research Corporation, Troy, Michigan.

Atrek, E., Agarwal, B. (1992), "Shape Optimization of Structural Design", Scientific Excellence in Supercomputing - The IBM 1990 Contest Prize Papers, (K.R. Billingsley, H.U. Brown III, E. Derohanes, Editors), Baldwin Press, University of Georgia, Athens, Georgia, pp. 311-351.

Atrek, E., Kodali, R. (1989), "Optimum Design of Continuum Structures with SHAPE", CAD/CAM Robotics and Factories of the Future, (B. Prasad, Editor), Vol. 2, (Proceedings of the 3rd International Conference: CARS and FOF '88), Springer Verlag, Berlin, pp. 11-15.

Bendsoe, M.P., Kikuchi, N. (1988), "Generating Optimal Topologies in Structural Design Using a Homogenization Method", Computer Methods in Applied Mechanics and Engineering, Vol. 71, pp. 197-224.

Bennett, J.A., Botkin, M.E. (1983), "Shape Optimization of Two-Dimensional Structures with Geometric Problem Description and Adaptive Mesh Refinement", AIAA/ASME/ASCE/AHS 24th Structures Structural Dynamics and Materials Conference (Part 1), Lake Tahoe, Nevada, pp. 422-431.

Botkin, M.E. (1981), "Shape Optimization of Plate and Shell Structures", AIAA/ASME/ASCE/AHS 22nd Structures Structural Dynamics and Materials Conference(Part 1), Atlanta, Georgia, pp. 242-249.

Cristescu, M., El-Yafi, F. (1986), "Interactive Shape Optimization", Computer Aided Optimal Design: Structural and Mechanical Systems, (NATO/NASA/NSF/USAF Advanced Study Institute, Troia, Portugal, June 29-July 11, 1986), (C.A. Mota Soarez, Editor), Vol. 2, Center for Mechanics and Materials of the Technical University of Lisbon, pp. 262-271.

Fleury, C. (1986), "Shape Optimal Design by the Convex Linearization Method", The Optimum Shape: Automated Structural Design, (J.A. Bennett, M.E. Botkin, Editors), Plenum Press, New York, pp. 297-320.

Henkel, F.-O., Schumann-Luck, A., Atrek, E. (1991), "Formoptimierung mit NISA-SHAPE", XX. International Finite Element Congress (FEM '91), Baden-Baden, Federal Republic of Germany.

Hou, G.J.W., Sheen, J.S., Chuang, C.H. (1990), "Shape Sensitivity Analysis and Design Optimization of Linear, Thermoelastic Solids", AIAA/ASME/ASCE/AHS/ASC 31st Structures Structural Dynamics and Materials Conference, Long Beach, California, (AIAA Paper No. 90-1012).

Imam, M.H. (1982), "Three-Dimensional Shape Optimization", International Journal for Numerical Methods in Engineering, Vol. 18, pp. 661-673.

Irons, B.M. (1970), "A Frontal Solution Program for Finite Element Analysis", International Journal for Numerical Methods in Engineering, Vol. 2, pp. 5-32.

Kikuchi, N., Suzuki, K. (1991), "Application of the Homogenization Method to Computational Mechanics and Optimum Structural Design: Structural Optimization of a Linearly Elastic Structure Using the Homogenization Method", IBM CSM Short Course, Rancho Mirage, California.

Kuhn, H.W., Tucker, A.W. (1951), "Nonlinear Programming", Proceedings of the Second Berkeley Symposium On Mathematical Statistics and Probability, (J. Neyman, Editor), July 31-August 12, 1950, University of California Press, Berkeley, pp. 481-492.

Kumar, V., German, M.D., Lee, S.-J. (1989), "A Geometry-Based 2-Dimensional Shape Optimization Methodology and a Software System with Applications", CAD/CAM Robotics and Factories of the Future, (B. Prasad, Editor), Vol. 2, Springer Verlag, Berlin, pp. 5-10.

Maier, G. (1973), "Limit Design in the Absence of a Given Layout: A Finite Element Zero-One Programming Problem", Optimum Structural Design-Theory and Applications, (R.H. Gallagher, O.C. Zienkiewicz, Editors), John Wiley & Sons, London, pp. 223-239.

Rodriguez-Velazquez, J., Seireg, A.A. (1985), "Optimizing the Shapes of Structures Via a Rule-Based Computer Program", Computers in Mechanical Engineering, Vol. 4, No. 1, pp. 20-28.

Rodriguez(-Velazquez), J., Seireg, A. (A.) (1988), "A Geometric Rule-Based Methodology for Shape Synthesis: 2-D Cases", Proceedings 1988 ASME International Computers in Engineering Conference and Exposition, San Francisco, California, Vol. 2, pp. 35-41.

Author's address:

Doz. Dr. Erdal Atrek
DaRC International
873 Bridge Park
Troy, MI 48098, USA
(Previously at Engineering Mechanics Research Corporation,
1607 E. Big Beaver Rd., Troy, MI, 48083 USA)

STARS
Mathematical Foundations

P Bartholomew
and
Sarah Vinson

Abstract. This chapter summarises the theoretical basis of the structural design program STARS. Various methods available within the code are described, with particular emphasis on the Newton algorithm. This is shown to be closely related to the use of an optimisation model with a quadratic approximation to the objective function and a linear approximation to the constraints. Particular reference is also made to problems such as non-differentiability and inaccuracy of sensitivities based on truncated model expansions which occur is dynamics applications.

1. INTRODUCTION

1.1 General Outline

The last decade has seen considerable expansion in the role of computer based structural design. This is evident both from the development of CAD/CAM systems and from the increased use of Finite Element analysis in design. More recently this process has extended to the introduction of optimisation techniques within the process of structural design.

STARS (STructural Analysis and Redesign System) is a modular computer program developed by the DRA and EDS-Scicon for the automated design of minimum weight structures subject to a variety of behavioural constraints. The most effective use of these techniques in a design environment requires their presentation within an integrated system. STARS is combined with other software in such a way that the inherent complexity does not obscure the overall aim of designing an optimised structure.

The creation of an optimisation or automated design program is straightforward in concept, but practical considerations make the task complex. Over the past two decades a number of individual structural optimisation methods have been proposed (see Morris 1982 and Atrek *et al* 1984) but each tends to be effective for a different class of problem. A major requirement of STARS is that it should be sufficiently flexible to accommodate a variety of these optimisation methods so that a wide range of problems can be addressed.

1.2 Scope of the System

STARS is intended to aid the design of thin-walled structures by determining member thicknesses. While the methods are primarily aimed at optimising the membrane load carrying capability of the structure, its load-bearing and stability behaviour in bending are also addressed. Both the geometry and material properties form part of the problem definition. Where fibre-reinforced composite laminates are specified, however, only the membrane properties are optimised at present. The objective functions considered are linear combinations of component volumes, usually representing structural weight. Although the selection of objective function is significant, the satisfaction of behavioural constraints is considered to be far more important in practice.

To enable the definition of strength designed problems, upper and lower limits may be placed on the stress carried by beams, von-Mises stress is available as the failure criterion for plates, and Tsai-Hill and fibre strain criteria are used for composites. Stiffness design may be carried out in combination with strength requirements by placing limits on acceptable deformation at any point of the structure. These deformations may either be specified as single components of deflexion or as linear combinations of such.

In aircraft design a major requirement for stiffness arises from aeroelasticity considerations, and these effects may be included in the analysis performed by STARS. Static stiffness constraints may be introduced to limit the efficiency loss of control surfaces; analysis software for flutter and static divergence calculation has been developed and may be run in conjunction with STARS to provide an aeroelastic optimisation capability. It is also possible to place constraints on the natural frequencies of individual modes of vibration, in isolation or in combination with any other constraint or analysis type.

An important feature of STARS is the use of duality theory to provide a monitoring facility for the performance of numerical algorithms. This gives information for checking the progress of an optimisation method during a given computer run, indicates which constraints are active, and also permits monitoring of the overall convergence of the algorithm.

1.3 Software Organisation

STARS uses an open architecture database configuration and all modules are totally integrated. Due to the interdependence of the modules, STARS has a single central database or Common Exchange (CX) file which contains sufficient information to enable a restart of the optimisation from the last completed iteration. In addition, the user has the option to take a copy of this CX file at any point in the optimisation process which can then be used as both a backup and an initial starting point for a revised optimisation procedure. The flexibility of the system enables the user to schedule the STARS modules together in a logical sequence to suit his requirements.

Due to the open architecture of the system, new modules and site specific routines, implementing additional types of constraints for example, can be introduced through user-supplied routines.

1.4 CAE Integration, pre- and post-processing

The program is capable of interfacing with any of the commercially available structural analysis packages but, in addition, has a self-contained analysis module. The standard released product version of STARS has an interface to MSC NASTRAN supplied. An X-Y graph plotting programme can be used to monitor variations in weight, design variables and constraints as the optimisation progresses. Neutral files can be produced, for subsequent post-processing by PATRAN, FEMVIEW, FAMRESULT or I-DEAS, either by calling the interface module for execution, in batch, during the STARS analysis itself or alternatively by calling it in demand mode when the optimisation run has completed.

An aeroelastics module is available with STARS which was developed jointly by British Aerospace, Kingston and the DRA, Farnborough. This enables flutter, roll and divergence speed constraints to be handled and integrated with strength design optimisation constraints. The first phase of work aimed at including detail design methods as part of a multi-level optimisation scheme has recently been completed and plans have been formulated to extend this work further. The first phase involved the integration of a panel design program into the optimisation cycle.

1.5 Availability

STARS has been jointly developed by the DRA Farnborough and EDS-Scicon and is wholly owned by the UK Secretary of State for Defence. EDS-Scicon have the non-exclusive marketing rights to the software world-wide.

STARS is available for either lease or purchase on APOLLO, SUN, VAX, IBM, CRAY and ALLIANT hardware. Program documentation includes a User Manual, Applications Manual and Programmer's Manual.

2. OPTIMISATION MODEL

2.1 Mathematical Formulation

By the end of the 1960's there was a significant output of research work (see Schmitt and Pope 1973), which recognised structural design as an application of optimisation theory. The aim is to set up a general computer program incorporating appropriate methods for solving the problem

Minimise the structural weight $\quad F(A)$

subject to constraints $\quad g_i(A) \geq 0 \quad (i=1,...m)$, (1)

where A represents properties of the structural elements to be used as design freedoms. In theory this was achievable directly by employing the techniques of mathematical programming then available and allowing calls to Finite Element analysis for function evaluation. In practise, for any but the smallest of problems, an excessive demand for function evaluations, combined with the relatively high cost of each Finite Element analysis, tended to make the approach impractical. As a result there was substantial interest in the more engineering intuitive techniques such as Stress-ratioing and the Optimality Criterion method, and it was these which were more appropriate to industrial application at that time.

More recently it has become common practice to use the function evaluations and sensitivity calculations, derived from the finite element analysis, to create an Optimisation Model (as in Eschenauer 1989) and then use the optimisation model to supply information to the optimisation algorithm, thus separating the optimisation algorithm from the analysis function. Although this distinction exists in STARS, it is less apparent than in some more recent codes. At the time the main optimisation methods used by STARS were first developed, priority was given to interpreting the techniques of Stress-ratioing and the Optimality Criterion method in the context of mathematical optimisation theory, so that the assumptions or approximations which made them effective could then be exploited in the context of more conventional optimisation methods. Specifically, in the context of a statically-determinate structure, it may be shown that the use of reciprocal design variables can linearise common behavioural constraints on static stress or displacement. In such cases both the constraints and weight as an objective function are separable, substantially simplifying and optimisation.

In order to invoke as few function evaluations as possible, a Newton approach was employed, in which the approximations underlying the Fully Stressing and Optimality Criterion methods were used to provide a simple approximation to the Hessian matrix. The use of any such approximation implicitly defines an optimisation model, which in this instance takes the form of a separable quadratic objective function and a linear

STARS: mathematical foundations

approximation to the constraints. In STARS there is therefore a direct correspondence between the optimisation model and the solution method since the approximation can be minimised at a single step of the optimisation algorithm. This tends to mask the distinction between the two but, in principle, a wide variety of optimisation methods could be used to solve each of a sequence of local approximations to the design optimisation problem, with very little difference in computational efficiency.

More recently considerable effort has been devoted to generalising the optimisation model in order to incorporate non-linearity in the constraints resulting from the optimisation of plate bending structures and constraints arising from the introduction of dynamics and aeroelastic requirements.

2.2 Design variable linking and constraints

Although, in principle, no restriction need be placed on the form for $F(A)$ (several of the techniques described below can accommodate complicated forms for the objective function), a simple weight function is used in STARS. The variables A_k are taken to represent a structural dimension such as the cross-sectional area of a strut or the thickness of a panel. Since only rarely will one wish to vary every such quantity independently these element variables are linked to a reduced set of dimensionless design variables x_j ($j=1,...n$) using a Boolean matrix $\mathbf{B}$:

$$A_k = \sum_{j=1}^{n} a_k B_{kj} x_j^\alpha \tag{2}$$

where a_k is an initial reference area for each of the N elements. The weight is a separable function of x_j given by

$$f(x) = \sum_{j=1}^{n} w_j x_j^\alpha \tag{3}$$

where individual component weights w_j are given by

$$w_j = \sum_{k=1}^{N} \rho_k a_k l_k B_{kj} \tag{4}$$

and l_k represents the length of a strut or the surface area of a plate element, while ρ_k is its density.

In general terms, constraints are placed on the behavioural response of the structure. They may be non-linear functions of displacement and also may depend directly on the design variables. Where possible, the constraints are formulated in such a way that they at least scale linearly with design variables. Arguments based upon statically-determinate structures show that behavioural constraints on stress or displacement allowables are best modelled if the inverse design variables are employed, that is the exponent $\alpha=-1$ in equation (3). This improved representation of the constraint more than compensates for the loss of linearity in the objective function. The new structural optimisation problem:

Minimise
subject to
$$f(x) \equiv \sum_{j=1}^{n} w_j x_j^{-1}$$
$$g_i(x) \leq c_i, \quad (i = 1\ldots,m)$$
(5)

has an objective function which, although non-linear, is separable and convex. In addition to behavioural constraints, upper and lower bounds are placed on all design variables. Such gauge limits are introduced to help meet manufacturing, handling or other requirements not explicitly modelled in the analysis.

3. STRUCTURAL ANALYSIS

3.1 Finite Element Analysis

3.1.1 In-house DRA system. Since STARS is intended to interface with any standard analysis system at a user site, the provision of an analysis capability is not regarded as essential. However, its inclusion enables STARS to operate independently of other software and, in practice, its use is the strongly preferred option during code development. The range of elements is restricted to the following:

- a simple rod element capable of carrying end-load;
- a standard cubic displacement beam element which includes the effects of shear flexibility and offsets at nodes;
- the constant strain triangle (see Zienkiewicz 1977);
- an isoparametric quadrilateral membrane element (Zienkiewicz 1977) which uses four-point Gauss quadrature;
- a shear panel derived from the isoparametric quadrilateral by setting material properties other than shear stiffness to zero and reducing the quadrature to single point;
- a triangular plate-bending element, due to Allman (1976), is used for the optimisation of balanced honeycomb structures or solid-section plates;
- a point mass which is used either as a structural mass, such as that used for mass balance in dynamics problems, or to enable use of weight engineer's data as an alternative to using consistent structural mass.

The analysis module performs static, forced response and natural frequency analyses of the structure using the displacement finite element method. The stiffness matrix is stored using the "skyline" approach and in a linear statics analysis the equations are solved using an LDL^T decomposition. Formally, the analysis requires the solution of

$$\mathbf{K}\,\mathbf{u} = \mathbf{p} \qquad (6)$$

where the stiffness matrix $\mathbf{K}$ relates the vector of applied loads $\mathbf{p}$ to the vector of nodal displacements $\mathbf{u}$. The normal-mode calculation also requires calculation of the mass matrix, and the solution of the eigen-value equation

$$[\mathbf{K} - \omega_r^2 \mathbf{M}]\,\psi_r = 0 \qquad (7)$$

where ψ_r are the normal modes of vibration and ω_r are the corresponding natural frequencies. This eigen problem is actually solved in the form

$$\left[\mathbf{L}^{-1}\mathbf{D}^{-\frac{1}{2}}\mathbf{M}\,\mathbf{D}^{-\frac{1}{2}}\mathbf{L}^{-T}\right]\psi_r^{(p+1)} = \left(\frac{1}{\omega_r^2}\right)\psi_r^{(p)} \qquad (8)$$

using simultaneous iteration, following Corr and Jennings (1976).

In the forced response calculation, the structure is assumed to be subject to sinusoidally varying forced vibration at circular frequency Ω. The starting point is the general equation of motion for a dynamic system with viscous damping

$$\mathbf{M}\frac{d^2\mathbf{u}}{dt^2} + \mathbf{C}\frac{d\mathbf{u}}{dt} + \mathbf{K}\,\mathbf{u} = \mathbf{p}(t). \qquad (9)$$

For this form of excitation, the response will have in-phase and out-of-phase components $\mathbf{u}_1$ and $\mathbf{u}_2$ such that $\mathbf{u}(t)=\mathbf{u}_1\cos(\Omega t)+\mathbf{u}_2\sin(\Omega t)$, otherwise written $\mathbf{u}(t)=\{\mathbf{u}_1+i\mathbf{u}_2\}\exp(-i\Omega t)$. Equations governing the forced response are then obtained in the form

$$\begin{bmatrix} \mathbf{K}-\Omega^2\mathbf{M} & \Omega\mathbf{C} \\ -\Omega\mathbf{C} & \mathbf{K}-\Omega^2\mathbf{M} \end{bmatrix}\begin{Bmatrix}\mathbf{u}_1\\ \mathbf{u}_2\end{Bmatrix} = \begin{Bmatrix}\mathbf{p}_1\\ \mathbf{p}_2\end{Bmatrix}. \qquad (10)$$

The presence of the damping term $\Omega\mathbf{C}$ is critical if the forcing is near resonance, when $[\mathbf{K} - \Omega^2\mathbf{M}]$ is near-singular. While STARS will accept this set of equations as the basis for structural optimisation, the internal analysis capability assumes zero damping.

The loads themselves may be weakly dependent on displacement for static aerodynamic calculations. That is, an externally calculated Aerodynamic Influence Coefficient matrix $\mathbf{A}$ may also be introduced to provide additional loads generated by the displacement of the structure. The equation

$$\mathbf{K}\,\mathbf{u}^{(p+1)} = \mathbf{p} + \mathbf{A}\,\mathbf{u}^{(p)} \qquad (11)$$

is solved iteratively, and converges rapidly to to solution provided the aerodynamic influence is small relative to the structural stiffness.

Further dynamic analysis also takes place in the STARS aeroelastic modules. The flutter equation

$$\mathbf{M}\frac{d^2\mathbf{u}}{dt^2} + [\mathbf{B}V + \mathbf{C}]\frac{d\mathbf{u}}{dt} + \left[\mathbf{A}V^2 + \mathbf{K}\right]\mathbf{u} = 0 \qquad (12)$$

is first reduced to modal co-ordinates using a truncated set of zero-velocity modes, then solved as a complex eigen-value problem

$$\left[\overline{\mathbf{M}}\lambda_r^2 + (\overline{\mathbf{B}}V + \overline{\mathbf{C}})\lambda_r + (\overline{\mathbf{A}}V^2 + \overline{\mathbf{K}})\right]\phi_r(V) = 0 \qquad (13)$$

using inverse iteration to track modes at increasing velocity. Any complex eigenvalues λ_r occur in complex conjugate pairs of the form $\gamma_r \pm i\omega_r$ where γ_r is the modal decay and ω_r is the corresponding frequency.

The calculation of design sensitivities for statics, forced response and statics with aeroelastic increments of load all require re-entry with further load cases, so this is catered for specifically and is discussed in section 4. By comparison with commercially available finite element packages this analysis capability is limited, but this makes it rapid and economical in execution while providing the analysis capability relevant to the optimisation process.

3.1.2 NASTRAN. Each external analysis module usually requires two translator modules to be added to the package. While these modules are unseen by the user, they are vital for communication. In the case of Nastran, the interface works by calling Nastran itself to read and store details of the entire model as created by the user. STARS retrieves the data it requires via Nastran's OUTPUT2 files and subsequently it creates INPUT2 files to modify data in the Nastran database.

The first step is for the module NSIN (Nastran Structural INput) to read the Case Control Section of the Nastran deck submitted by the user and store the load case details on the CX file. NSIN then creates a new Nastran deck which consists of Executive Control and Case Control Sections for a Nastran SOL 60 datacheck. The Bulk Data Section is copied from the user's Nastran deck to the NSIN datacheck deck.

This deck is submitted to MSC/NASTRAN for data checking and for Nastran to create two OUTPUT2 binary files which contain all the grid point and structural data required by STARS. The data is stored on the CX file and a checklist written to the STARS results file.

To perform a finite element analysis using Nastran, the STARS module NAST is used to read the structural data from the CX file and identify the elements, associated with design variables, whose properties are changed during optimisation. It then reads the necessary data blocks from the Nastran output files and modifies the property numbers belonging to these elements. The modified data blocks are written to the INPUT2 binary file. Further information is also read from the output files and, after the INPUT2 file has been created, a Nastran deck is produced. The revised structure is then submitted to Nastran for analysis. Nastran then creates an output file containing the displacements for each node (and, on the initial iteration, two further files containing the element stiffness matrices). Finally the new output files are read by NAST and the results are stored on the CX file.

3.1.3 Approximate Analysis.

In the latter stages of a structural optimisation the design changes will be relatively small and it is possible to approximate the structural response on the basis of a previous full analysis. The design change will yield a modified stiffness matrix $[K+\Delta K]$ and the resulting displacements are assumed known to first order $u^{(p)}$. An iterative update similar to (11) is performed, based on

$$K\, u^{(p+1)} = p - \Delta K\, u^{(p)}. \tag{14}$$

The equations involve the use of a stiffness matrix which has already been factored but with a new load vector. In practice, as only a single step is made for each call for an approximate analysis, these iterations are normally carried out in parallel with the design iterations.

An approximate analysis capability has also been implemented within STARS to re-approximate the eigenvectors of the modified system for dynamics analysis (reported in Bartholomew 1991) or as part of the calculation of flutter and divergence speeds. It uses the previously calculated modes as basis vectors in order to calculate new modes within that sub-space. The stiffness and mass matrices, corresponding to the perturbed system, are first reduced to modal co-ordinates using the truncated set of eigen-vectors

$$\Psi = [\psi_1 \ldots, \psi_m, \vartheta_r], \tag{15}$$

giving
$$[\overline{K} - \omega_r^2 \overline{M}]\, \phi_r = 0. \tag{16}$$

where
$$\overline{K} = \Psi^T [K + \Delta K] \Psi$$
$$\overline{M} = \Psi^T [M + \Delta M] \Psi.$$

The reduced system of equations is readily solved for the complete set of eigen-vectors ϕ_r, $r=1,\ldots m$ to give an approximation to the eigenvectors of the perturbed system in the form

$$\{\psi_r\}_{x+\Delta x} = [\Psi]\, \phi_r. \tag{17}$$

It was found that the use of the truncated set of eigen-vectors gave rise to substantial errors due to the lack of any contribution from the higher order modes, so the basis vectors in equation (15) are augmented by a Ritz vector which is the solution of

$$\mathbf{K}\,\vartheta_r = \left[\frac{d\mathbf{K}}{ds} - \omega_r^2 \frac{d\mathbf{M}}{ds}\right]\psi_r + \frac{d(\omega_r^2)}{ds}\mathbf{M}\psi_r\,. \tag{18}$$

This solution uses existing factors for the stiffness matrix $\mathbf{K}$, available from the associated statics analysis, to give a computationally effective procedure.

The economy of approximate analysis is such that its use can lead to substantial savings in computer time. In addition it is used to provide estimates of higher order derivatives of constraints in order to improve the performance of optimisation algorithms. Approximate analysis in STARS is based on the internal Analysis capability or on NASTRAN.

3.2 Static Stiffness Requirements

The simplest form of constraint used by STARS is that placed on a single component of the displacement at a prescribed node. A slightly more general form of constraint is placed on a linear combination of displacements at different nodes.

$$g(x) = \mathbf{e}^T\mathbf{u} \leq c \tag{19}$$

where $\mathbf{e}$ is an arbitrary vector and $\mathbf{u}$ is the solution vector of displacements for the required load case. Such a constraint may be used for purposes as diverse as limiting the angle of twist in a vehicle chassis, increasing the stiffness of modes of vibration, reducing the relative movement between parts of a structure and imposing aeroelastic efficiency constraints.

3.3 Static Strength Requirements

3.3.1 Strain. These constraints are treated as further generalisations of the previous functions of displacement. The primary application of strain constraints is to the design of structures using composite materials. Individual components of the strain matrix may be constrained, typically with reference to fibre strain, transverse strain and shear strain allowables of a lamina. The membrane strains ε_0 are calculated at the element centroid

$$\varepsilon_0 = \mathbf{B}\,\mathbf{u} \tag{20}$$

using a strain/displacement matrix $\mathbf{B}$ from the constant strain triangle or isoparametric quadrilateral as appropriate. This matrix is also required, as shown in section 4.3, to calculate sensitivities consistent with the constraint.

The contributions to stresses and strains resulting from bending action are calculated from bending moments derived by application of the principle of minimum complementary energy, giving a "best fit" to prescribed displacements on the element boundary. That is

$$M = F^{-1} G T u \qquad (21)$$

where Tu are the generalised displacements, GTu is the complementary work corresponding to each of the constant moment stress fields and F is the flexibility matrix. The element curvatures then follow from

$$\kappa = D^{-1} M \qquad (22)$$

where D is the bending stiffness matrix. The strain at a general stress recovery point is a combination of membrane and bending contributions, *viz*:

$$\{\varepsilon\} = \{\varepsilon_0\} + \{\kappa\}h \qquad (23)$$

where h is the offset from the neutral axis to the stress recovery point.

3.3.2 Stresses.
As for strains, the stresses used as constraints are calculated from displacements within STARS, even though stresses may be available from external analysis programs. The calculation follows from the strain calculations above:

$$\sigma = Q \varepsilon \qquad (24)$$

where Q is the stress/strain matrix. Constraining a single component of stress is appropriate for the simplest elements but the von Mises equivalent stress or the Tsai Wu failure criteria is more appropriate for plate elements with metallic or composite material properties respectively. The von Mises stress is given by

$$\sigma_{VM}^2 = 1/2 \, \{\sigma\}^T [V] \{\sigma\} \qquad (25\ a)$$

where

$$[V] = \begin{bmatrix} 2 & -1 & 0 \\ -1 & 2 & 0 \\ 0 & 0 & 6 \end{bmatrix}. \qquad (25\ b)$$

The Tsai-Wu constraint is also a quadratic form but in this instance allowance is made for the differing stress allowables appropriate to composite material, *viz*

$$\frac{\sigma_1^2}{\bar{\sigma}_1 \bar{\sigma}_{1c}} - \frac{\sigma_1 \sigma_2}{(\bar{\sigma}_1 \bar{\sigma}_{1c} \bar{\sigma}_2 \bar{\sigma}_{2c})^{1/2}} + \frac{\sigma_2^2}{\bar{\sigma}_2 \bar{\sigma}_{2c}} + \frac{\tau_{12}^2}{\bar{\tau}_{12}^2} + \left(\frac{1}{\bar{\sigma}_{1t}} - \frac{1}{\bar{\sigma}_{1c}}\right)\sigma_1 + \left(\frac{1}{\bar{\sigma}_{2t}} - \frac{1}{\bar{\sigma}_{2c}}\right)\sigma_2 \leq 1 \qquad (26)$$

where the suffices 1 and 2 denote the ply fibre direction and transverse direction respectively and the bar denotes a failure stress. The suffices t and c indicate tensile and compressive failure modes respectively. For use in STARS, the constraint is transformed, by solving (26) as a quadratic function of t, to yield a form which is approximately linear in inverse variables, namely

$$g \equiv \frac{1}{2}\left\{\left(\frac{1}{\bar{\sigma}_{1t}}-\frac{1}{\bar{\sigma}_{1c}}\right)\sigma_1 + \left(\frac{1}{\bar{\sigma}_{2t}}-\frac{1}{\bar{\sigma}_{2c}}\right)\sigma_2\right\} \qquad (27)$$

$$\pm \left\{\frac{1}{4}\left[\left(\frac{1}{\bar{\sigma}_{1t}}-\frac{1}{\bar{\sigma}_{1c}}\right)\sigma_1 + \left(\frac{1}{\bar{\sigma}_{2t}}-\frac{1}{\bar{\sigma}_{2c}}\right)\sigma_2\right]^2 + \left(\frac{\sigma_1^2}{\bar{\sigma}_{1t}\bar{\sigma}_{1c}} - \frac{\sigma_1\sigma_2}{(\bar{\sigma}_{1t}\bar{\sigma}_{1c}\bar{\sigma}_{2t}\bar{\sigma}_{2c})^{1/2}} + \frac{\sigma_2^2}{\bar{\sigma}_{2t}\bar{\sigma}_{2c}} + \frac{\tau_{12}^2}{\bar{\tau}_{12}^2}\right)\right\}^{\frac{1}{2}}$$

3.4 Panel Buckling

A constraint involving the direct and shear resultants N_x, N_{xy} in the form

$$-a_1\frac{N_x}{t^3} + a_2^2\left(\frac{N_{xy}}{t^3}\right)^2 \leq 1 \qquad (28)$$

is applicable to the local buckling of simple or stiffened panels under a variety of edge support conditions. The user must specify the compressive and shear buckling stresses σ_{crit}, τ_{crit} for a selected reference panel thickness t_{ref}, and so implicitly defines the coefficients a_1 and a_2, viz

$$a_1 = -\frac{t_{ref}^2}{\bar{\sigma}_{crit}}, \qquad a_2 = \frac{t_{ref}^2}{\bar{\tau}_{crit}}. \qquad (29)$$

For use in STARS, the constraint is transformed, by solving (28) as a quadratic function of t^3, to yield a form which is approximately linear in inverse variables, namely

$$g(x) \equiv \left\{\frac{t_{ref}^2}{2t^2}\right\}^{\frac{1}{3}}\left\{\frac{\sigma_x}{\bar{\sigma}_{crit}} + \left(\frac{\sigma_x^2}{\bar{\sigma}_{crit}^2} + \frac{4\tau_{xy}^2}{\bar{\tau}_{crit}^2}\right)^{\frac{1}{2}}\right\}^{\frac{1}{3}} \leq 1. \qquad (30)$$

In this form, even the stress-ratio method may be applied with success.

3.5 Natural Frequency

Lower bounds may be prescribed for the frequencies of specified modes, viz

$$\omega_r \geq \bar{\omega}_r. \qquad (31)$$

It is easier to satisfy such a constraint in the presence of significant non-structural mass. In that case, increasing structural stiffness will raise the natural frequencies, so giving a better chance of achieving a feasible design. In STARS the constraint is applied in the form

$$g(x) \equiv \frac{\bar{\omega}_r^2}{\omega_r^2} \leq 1 \qquad (32)$$

where ω_r is given by

$$\omega_r^2 \equiv \frac{\psi_r^T K \psi_r}{\psi_r^T M \psi_r}, \qquad (33)$$

showing the inverse dependence of the constraint (32) on K explicitly. Where two or more natural frequencies coincide, often as a result of optimisation against a prescribed common lower limit, the constraints are nondifferentiable. In this case the design step is restricted by a further interaction constraint (see Bartholomew and Pitcher 1984),

$$g(x) \equiv \psi_r^T [K - \omega_{r,s}^2 M] \psi_s = 0, \qquad (34)$$

which is satisfied as an identity at the current design.

A characteristic of optimising a structure against such requirements is that the mode shapes corresponding to a particular constraint may change drastically during the optimisation, as shown in Bartholomew (1990). An alternative approach, which concentrates on a specific physical mode of the original structure, is obtained by applying a static load $M\psi_r$ and requiring the resulting displacement field to satisfy $\psi_r^T M u_r \leq 1/\omega_r^2$. This constraint is subsequently treated as a generalised displacement constraint, with the sole difference that it applies only to the single load case.

3.6 Forced response

The constraint is of the form

$$g \equiv \{u_1^2 + u_2^2\}^{\frac{1}{2}} \leq c, \qquad (35)$$

where u_1 and u_2 are in-phase and out-of-phase components of displacement at a particular node, given by $u_1 = \hat{e}^T u_1$, $u_2 = \hat{e}^T u_2$.

3.7 Aeroelastic Requirements

3.7.1 Static Aeroelastic Constraints.
A further form of the generalised displacement constraint arises from static aeroelastic considerations, where it may be necessary to bound the increments of aerodynamic load due to the elastic deformation. In this case an efficiency constraint may be placed on forces or moments on the flexible structure relative to values calculated for the rigid component. The constraint is of the form

$$b^T p + b^T A u \geq \eta b^T p \qquad (36)$$

where b is a vector chosen to provide an appropriate linear combination of the aerodynamic forces. Equation (36) may be rewritten as

$$-(1-\eta) \leq \frac{\mathbf{b}^T \mathbf{A} \mathbf{u}}{\mathbf{b}^T \mathbf{p}} \tag{37}$$

and hence the coefficients used in the generalised displacement constraint are given by

$$\mathbf{e} = \frac{\mathbf{A}^T \mathbf{b}}{\mathbf{b}^T \mathbf{p}} . \tag{38}$$

Constraints may also be applied to initial and asymptotic roll rate efficiencies.

3.7.2 Flutter and Divergence.

The critical flutter and divergence speeds are determined by tracking the eigenvectors λ over a range of airspeeds. The constraints are imposed as upper bounds on the inverse of the critical velocities, thus

$$\frac{\overline{V}}{V_d} \leq 1, \quad \frac{\overline{V}}{V_f} \leq 1. \tag{39}$$

In order to address hump modes it is also possible to constrain modal decay at a range of fixed velocity stations. To improve the smoothness of the constrained quantities at divergence, the constraints are imposed on the sum and product of pairs of roots

$$\tfrac{1}{2}(\lambda_r + \lambda_s) \leq 0, \quad -\lambda_r \lambda_s \leq 0, \tag{40}$$

where equality of the first constraint implies zero modal decay γ_r and the onset of flutter, whereas equality of the second constraint implies a zero frequency and the onset of divergence.

4. SENSITIVITY ANALYSIS

4.1 General Strategy

The general class of optimisation methods used by STARS requires that the derivatives of the constraints, $\partial g_i/\partial x_j$, are supplied by the main program for each member of the active constraint set. In order to describe how these derivatives are obtained it is convenient to refer back to equations in the analysis section 3.

Firstly, noting equation (2), the derivatives of any constraint g_i with respect to the design variable x_j is given in terms of derivatives with respect to A_k by

$$\frac{\partial g_i}{\partial x_j} = \alpha \, x_j^{(\alpha-1)} \left[\sum_{k=1}^{N} a_k B_{kj} \frac{\partial g_i}{\partial A_k} \right] . \tag{41}$$

Such a summation of contributions due to the variation of elements controlled by a given design variable is carried out for all constraint types.

The requirement for these derivatives resolves into a need for the analysis module to supply stresses and displacements under the action of the applied loads, and then permit re-entry for analysis under the action of further loads applied in accordance with the type and number of active constraints.

4.2 Displacements

To establish the sensitivity of response quantities to structural change, the governing equation (6) is differentiated to give

$$\frac{\partial \mathbf{K}}{\partial A_k} \mathbf{u} + \mathbf{K} \frac{\partial \mathbf{u}}{\partial A_k} = \frac{\partial \mathbf{p}}{\partial A_k}. \tag{42}$$

Thus, assuming $\partial \mathbf{p}/\partial A_k = 0$, the sensitivity of the displacement vector is

$$\frac{\partial \mathbf{u}}{\partial A_k} = - \mathbf{K}^{-1} \frac{\partial \mathbf{K}}{\partial A_k} \mathbf{u}. \tag{43}$$

To calculate sensitivities from this expression by the direct method requires further solution of the system equations with load vectors representing the out-of-balance loading for each design variable and under each of the original load cases. In STARS, the sensitivity calculation is performed for each particular constraint by a method which later became known as the adjoint method. When the derivative of a specified nodal displacement $u = \hat{\mathbf{e}}^T \mathbf{u}$ is required, it may be written in the following form:

$$\frac{\partial u}{\partial A_k} \equiv \hat{\mathbf{e}}^T \frac{\partial \mathbf{u}}{\partial A_k} = - \hat{\mathbf{e}}^T \mathbf{K}^{-1} \frac{\partial \mathbf{K}}{\partial A_k} \mathbf{u}$$
$$= - \mathbf{v}^T \frac{\partial \mathbf{K}}{\partial A_k} \mathbf{u}, \tag{44}$$

in which only one further solution of the system equations is required. The additional load case is a unit load $\hat{\mathbf{e}}$, applied at the point corresponding to the designated displacement u, and its solution is the adjoint vector $\mathbf{v}$. The sensitivity of the generalised displacement constraint is the same except that the adjoint load is no longer a unit vector. The matrix product (44), although written here in terms of global quantities, is actually carried out using local element freedoms, thereby vastly reducing the dimensionality of the matrix algebra.

Because displacement finite elements are always used in STARS, any other derivatives which may be required, *eg* stress derivatives, are obtained from the expression for $\partial \mathbf{u}/\partial A_k$ by additional matrix algebra involving, for example, the material constitutive relations to obtain stress derivatives.

4.3 Static strength

4.3.1 Strain. Each component of strain is simply a linear combination of displacements involving the strain/displacement matrix **B**, in equation (20), giving

$$\frac{\partial \epsilon_0}{\partial A_k} = \mathbf{B}\, \frac{\partial \mathbf{u}}{\partial A_k}$$
$$= -\mathbf{B}\, \mathbf{K}^{-1}\, \frac{\partial \mathbf{K}}{\partial A_k}\, \mathbf{u}\, . \tag{45}$$

Thus the additional load cases required are the columns of $\mathbf{B}^T$, and the adjoint displacement **v** satisfies

$$\mathbf{K}\,\mathbf{v} = \mathbf{B}^T \hat{\mathbf{e}}\, , \tag{46}$$

where $\{\hat{\mathbf{e}}\}$ is a unit vector selecting a particular component of the strain vector.

4.3.2 Stress. The calculation of sensitivities of single components of stress, especially for rods, is very similar to the calculation for strains but with an extra term in the adjoint load, requiring solution of

$$\mathbf{K}\,\mathbf{v} = \mathbf{B}^T \mathbf{Q}\, \hat{\mathbf{e}} \tag{47}$$

where **Q** is the stress/strain matrix of equation (24).

Failure criteria based on the von Mises equivalent stress or the Tsai-Wu condition no longer reduce to linear combinations of displacement and the adjoint loads themselves become implicit functions of the design variables. The sensitivity for von Mises stress, (25), is given in terms of the derivatives of stress components by

$$\frac{\partial \sigma_{VM}}{\partial A_k} = \frac{1}{2\sigma_{VM}} \sigma^T \mathbf{V}\, \frac{\partial \sigma}{\partial A_k} \tag{48}$$

requiring solution of

$$\mathbf{K}\,\mathbf{v} = \frac{1}{2\sigma_{VM}} \mathbf{B}^T \mathbf{Q} \mathbf{V}\, \sigma\, . \tag{49}$$

Derivatives of the Tsai-Wu constraint, (26), are derived in a similar manner.

4.4 Panel Buckling

The local buckling constraint (30) differs from the previous forms of constraint in so far as it contains explicit as well as implicit functions of the thickness t. Thus, the derivative

$$\frac{\partial g}{\partial t_k} = -\frac{2g}{3t_k} + \left\{\frac{t_{ref}^2}{2t_k^2}\right\}^{\frac{1}{3}} \frac{\partial}{\partial t_k}\left\{\frac{\sigma_x}{\sigma_{crit}} + \left(\frac{\sigma_x^2}{\sigma_{crit}^2} + \frac{4\tau_{xy}^2}{\tau_{crit}^2}\right)^{\frac{1}{2}}\right\}^{\frac{1}{3}} \quad (50)$$

contains the additional term $-2g/3t_k$ which appears when differentiating with respect to the design variable controlling thickness of the panel to which the buckling constraint is applied. The remaining terms are calculated from an adjoint load as for the constraints above.

4.5 Natural Frequency

Design sensitivities for the natural frequencies are based on Rayleigh quotients and do not require re-entry to the analysis modules. The formula used, which may be obtained by differentiating (7), is

$$\frac{\partial(\omega_r^2)}{\partial A_k} = \frac{\psi_r^T\left[\frac{\partial \mathbf{K}}{\partial A_k} - \omega_r^2 \frac{\partial \mathbf{M}}{\partial A_k}\right]\psi_r}{\psi_r^T \mathbf{M} \psi_r}. \quad (51)$$

Where two or more natural frequencies coincide, the eigen-vectors are not uniquely determined, so applying this equation gives arbitrary results; the constraints being nondifferentiable at the current design point. The additional interaction constraint (34) is then used, whose sensitivity is given by

$$\frac{\partial g}{\partial A_k} = \psi_r^T\left[\frac{\partial \mathbf{K}}{\partial A_k} - \omega_{r,s}^2 \frac{\partial \mathbf{M}}{\partial A_k}\right]\psi_s. \quad (52)$$

This constraint serves to restrict design changes to directions in which the current eigenvectors remain unaltered and (51) is valid.

4.6 Forced Response

Differentiating the governing equations (10) with respect to an element cross-sectional area or thickness A_k, gives

$$\begin{bmatrix} K-\Omega^2 M & \Omega C \\ -\Omega C & K-\Omega^2 M \end{bmatrix} \begin{Bmatrix} \frac{\partial u_1}{\partial A_k} \\ \frac{\partial u_2}{\partial A_k} \end{Bmatrix} = -\begin{bmatrix} \frac{\partial K}{\partial A_k} - \Omega^2 \frac{\partial M}{\partial A_k} & \Omega \frac{\partial C}{\partial A_k} \\ -\Omega \frac{\partial C}{\partial A_k} & \frac{\partial K}{\partial A_k} - \Omega^2 \frac{\partial M}{\partial A_k} \end{bmatrix} \begin{Bmatrix} u_1 \\ u_2 \end{Bmatrix}. \tag{53}$$

The calculation of derivatives by the direct method again requires solution of the same set of governing equations, but with further load vectors. The adjoint approach requires the derivative of equation (35) to be expressed in the form

$$\frac{\partial g}{\partial A_k} = \frac{1}{g}\begin{pmatrix} u_1 \hat{e}^T & u_2 \hat{e}^T \end{pmatrix} \begin{Bmatrix} \frac{\partial u_1}{\partial A_k} \\ \frac{\partial u_2}{\partial A_k} \end{Bmatrix}. \tag{54}$$

Putting these equations together gives

$$\frac{\partial g}{\partial A_k} = -\begin{pmatrix} v_1^T & v_2^T \end{pmatrix} \begin{bmatrix} \frac{\partial K}{\partial A_k} - \Omega^2 \frac{\partial M}{\partial A_k} & \Omega \frac{\partial C}{\partial A_k} \\ -\Omega \frac{\partial C}{\partial A_k} & \frac{\partial K}{\partial A_k} - \Omega^2 \frac{\partial M}{\partial A_k} \end{bmatrix} \begin{Bmatrix} u_1 \\ u_2 \end{Bmatrix}, \tag{55}$$

where v_1 and v_2 are given by

$$\begin{pmatrix} v_1^T & v_2^T \end{pmatrix} = \frac{1}{g}\begin{pmatrix} u_1 \hat{e}^T & u_2 \hat{e}^T \end{pmatrix} \begin{bmatrix} K-\Omega^2 M & \Omega C \\ -\Omega C & K-\Omega^2 M \end{bmatrix}^{-1}. \tag{56}$$

This equation may be inverted to give a form similar to the original governing equation, but care must be taken when transposing the equation due to the asymmetry of the matrix. The form of the matrix operator may be completely restored by changing the sign of the out-of-phase part of the adjoint displacement, viz:

$$\begin{bmatrix} K-\Omega^2 M & \Omega C \\ -\Omega C & K-\Omega^2 M \end{bmatrix} \begin{Bmatrix} v_1 \\ -v_2 \end{Bmatrix} = \frac{1}{g} \begin{Bmatrix} u_1 \hat{e} \\ -u_2 \hat{e} \end{Bmatrix}. \tag{57}$$

Thus the first column of the adjoint load is $u_1 \hat{e}/g$ and the second column is $-u_2 \hat{e}/g$. When processing the adjoint displacements it is important to note the sign of v_2, which in the complex variable notation corresponds to the imaginary term of the complex conjugate vector $\{v_1 - iv_2\}$.

STARS: mathematical foundations

4.7 Aeroelastic Requirements

4.7.1 Static Aeroelastic Constraints.
The introduction of the aerodynamic influence matrix is complicated by the fact that the adjoint matrix, $[K - A^T]$, is neither banded nor symmetric. Thus, as for the analysis equation (11), an iterative process is preferred to the direct solution of the equations. Equation (44) is still valid but here $\{v\}$ is defined by

$$v^T [K - A] = e^T . \tag{58}$$

The iterative solution procedure used is

$$K v^{(p+1)} = p + A^T v^{(p)} , \tag{59}$$

which differs in form from (11) in so far as the matrix A is transposed.

4.7.2 Flutter and Divergence.
As with natural frequency, aeroelastic sensitivities can be found without knowledge of corresponding eigenvector sensitivities by premultiplying the partial derivative of the governing equation by the appropriate left eigenvector. In this case, working from equation (13) gives

$$\psi^T \left[2\overline{M} \lambda_r + \overline{B} V \right] \phi \frac{\partial \lambda_r}{\partial A_k} + \psi^T \left[\overline{B} \lambda_r + 2\overline{A} V \right] \phi \frac{\partial V}{\partial A_k} = - \psi^T \left[\frac{\partial \overline{M}}{\partial A_k} \lambda_r^2 + \frac{\partial \overline{K}}{\partial A_k} \right] \phi \tag{60}$$

This complex equation may be expressed in the alternative form of two real equations relating the sensitivities of the three quantities; modal decay γ_r, frequency ω_r and velocity V, where $\lambda_r = \gamma_r + i\omega_r$. The constraint sensitivity for flutter speed is determined by setting γ_r and $\partial \gamma_r / \partial A_k$ to zero and solving equation (60) for $\partial V_f / \partial A_k$ and $\partial \omega_r / \partial A_k$.

At divergence, equation (60) is substantially reduced to the form

$$\psi^T \left[\overline{A} \right] \phi \, 2V_d \frac{\partial V_d}{\partial A_k} = - \psi^T \left[\frac{\partial \overline{K}}{\partial A_k} \right] \phi \tag{61}$$

since λ_r and its derivatives are zero. The sensitivities of the constraints based on the sum and product of the eigenvalues at fixed velocity stations are evaluated from the eigenvalue sensitivities, which in turn are given by the solution of

$$\psi^T \left[2\overline{M} \lambda_r + \overline{B} V \right] \phi \frac{\partial \lambda_r}{\partial A_k} = - \psi^T \left[\frac{\partial \overline{M}}{\partial A_k} \lambda_r^2 + \frac{\partial \overline{K}}{\partial A_k} \right] \phi \tag{62}$$

where the fact that $\partial V / \partial A_k = 0$ is used to reduce equation (60) to the form shown.

5. OPTIMISATION ALGORITHMS

5.1 Kuhn Tucker necessary conditions

In order to facilitate the derivation of methods for structural design optimisation, it is easiest to reformulate the problem using Lagrange multipliers to enforce constraints. To do this it is convenient to define the associated Lagrangian function, which is given by the expression

$$L(x,\lambda) = f(x) + \sum_{i=1}^{m} \lambda_i (g_i(x) - c_i), \qquad (63)$$

The conditions which hold at the constrained minimum of the original problem (5) are also the conditions for a stationary point of the Lagrangian function (63). These equations are known as the Kuhn-Tucker optimality conditions,

$$\frac{\partial f}{\partial x_j} + \sum_{i=1}^{m} \lambda_i \frac{\partial g_i}{\partial x_j} = 0, \qquad (j = 1\ldots,n) \qquad (64\text{ a})$$

$$\left. \begin{array}{l} \lambda_i (g_i(x) - c_i) = 0 \\ g_i(x) \le c_i, \quad \lambda_i \ge 0 \end{array} \right\} \quad (i = 1\ldots,m). \qquad \begin{array}{l}(64\text{ b})\\ (64\text{ c})\end{array}$$

where (64a,b) may be obtained by differentiating the Lagrangian. If the problem is convex then these conditions are necessary and sufficient for the solution vector (x_j^*, λ_i^*) to represent a global optimising point, otherwise they simply define a stationary point.

5.2 Fully stressing algorithm

The assumption that every component of a structure may be sized so that it achieves its stress limit leads to simple traditional algorithms such as the stress ratio method. Such methods associate a specific constraint $\sigma_j(x)$ with each design variable x_j ; thus the number of active constraints is equal to the number of design variables. In terms of this selected set, the constraints equations (64 b,c) reduce to

$$\frac{\sigma(x_j)}{\overline{\sigma}} = 1, \quad \lambda_j > 0. \qquad (65)$$

The fully stressing algorithm ignores the first equation (64a) and simply seeks an optimum on the basis of constraint satisfaction. Estimates of the optimising values of the design variables x_j (k+1) are obtained from current values x_j (k) within STARS by approximating the relationship between the stress and the governing design variable, given by equations (23) and (24), by a power relationship

$$\sigma_j(x) \approx \sigma_j^{(k)} \left\{ \frac{x_j}{x_j^{(k)}} \right\}^\beta , \qquad (66)$$

thus giving

$$x_j^{(k+1)} = x_j^{(k)} \left\{ \frac{\bar{\sigma}_j}{\sigma_j^{(k)}} \right\}^{1/\beta} . \qquad (67)$$

A single step of the method is economical on computer time as it requires no derivative calculation, and is valuable in the early stages of strength critical design where it can give significant improvements in design. The introduction of the exponent β allows the approach to be used for plate bending problems, where the usual formula with $\beta=1$ would fail. The algorithm is stable but does not necessarily converge to the optimum. It may also be used profitably in combination with more general techniques where there is benefit in reducing the number of active constraints which are not simple bounds. Its underlying assumptions make the method inappropriate for other classes of problem such as vibration-limited design or stiffness-critical design.

5.3 Optimality Criterion

Conversely, the Optimality Criterion method is best applied to stiffness critical design, in which the optimisation is controlled by displacement constraints. In this situation the active constraints are likely to be few in number, therefore it is appropriate to concentrate on the first equation of (64a).

$$-\frac{w_j}{x_j^2} + \sum_{i=1}^m \lambda_i \frac{\partial g_i}{\partial x_j} = 0 \quad (j = 1, \ldots n) . \qquad (68)$$

If the constraint is assumed to be a separable function of the design variables, then this equation provides the basis for update formulae, described in Morris (1982), for determining the design as a function of λ_i. In the particular case of a statically determinate membrane structure, the constrained displacement will be a linear function of the reciprocal design variables and equation (68) is solved without iteration. The calculation is further simplified if only one constraint is active and the design may be scaled to give constraint satisfaction, in which case the above procedure will give the optimum solution immediately. For multiple active constraints, the envelope method is employed, in which it is assumed that each dual variable λ_i may be sized solely on the basis of information concerning the corresponding constraint. That is

$$\lambda_i^{(k+1)} = \lambda_i^{(k)} \left\{ \frac{g_i(x)}{c_i} \right\}^{2\alpha} \qquad (69)$$

where $\lambda_i\,(k+1)$ is the value of the dual variable to be used at the following iteration and the factor a is adjusted empirically to improve convergence or stabilise the method.

5.4 Newton method

5.4.1 Derivation of basic algorithm.
The method is based on the use of linear approximations to both the optimality conditions and the constraint equations. The resulting simultaneous linear equations

$$\sum_{k=1}^{n}\left[\frac{\partial^2 f}{\partial x_j \partial x_k} + \sum_{i=1}^{m}\lambda_i \frac{\partial^2 g_i}{\partial x_j \partial x_k}\right]\delta x_k + \sum_{i=1}^{m}\frac{\partial g_i}{\partial x_j}\delta\lambda_i = -\frac{\partial f}{\partial x_j} - \sum_{i=1}^{m}\lambda_i \frac{\partial g_i}{\partial x_j}, \quad (j=1\ldots,n) \qquad (70\text{ a})$$

$$\sum_{k=1}^{m}\frac{\partial g_i}{\partial x_k}\delta x_k = c_i - g_i, \qquad (i=1\ldots,m), \qquad (70\text{ b})$$

if solved as they stand, would give the increments δx_j, $\delta\lambda_i$ corresponding to a single step of the Newton method for the solution of the non-linear Kuhn Tucker conditions. Convergence would be quadratic, provided the starting design were sufficiently close to the optimum, though the method would be computationally intensive and it may be unstable when applied to non-convex problems.

However, the approximations which underlie both the optimality criterion method and stress-ratioing lead to substantial simplifications of the above equations. The assumption that introducing reciprocal variables linearises the constraints, reduces the problem to a standard linearly constrained problem of positive-definite quadratic programming. Moreover, since the term involving second derivatives of the constraints, underlined in equation (70a), is assumed to be zero, the Hessian matrix is in fact reduced to diagonal form. This reflects the separable nature of the structural optimisation problem which was exploited by the optimality criterion method.

Just as there each optimality equation may be used to size a single design variable, so here it gives a block structure for the $(n+m)$ system of equations such that increments of the primal variables δx_j may be eliminated algebraically, using (70a), in favour of the dual variables λ_i. Thus, denoting $\lambda_i^{(k)}+\delta\lambda_i$ by $\lambda_i^{(k+1)}$,

$$\delta x_j = \frac{x_j^{(k)}}{2} - \frac{x_j^{(k)3}}{2w_j}\sum_{i=1}^{m}\lambda_j^{(k+1)}\frac{\partial g_i}{\partial x_j} \qquad (j=1\ldots,n), \qquad (71)$$

which, substituted into equation (70 b), gives

$$\sum_{p=1}^{m}\left[\sum_{j=1}^{m}\frac{\partial g_i}{\partial x_j}\frac{x_j^{(k)3}}{2w_j}\frac{\partial g_p}{\partial x_j}\right]\lambda_p^{(k+1)} = (g_i - c_i) + \sum_{j=1}^{m}\frac{x_j^{(k)}}{2}\frac{\partial g_i}{\partial x_j} \qquad (i=1\ldots,m). \qquad (72)$$

Then a numerical solution is only required for a system of m equations in m unknowns, where m is here the number of *active* constraints. Stiffness constrained problems tend to be characterised by relatively small active constraint sets, leaving a relatively small set of equations to be solved, and importantly this does not increase in size as more design variables are introduced.

It may be noted in passing that this direct approach to the solution of equation (70) is not normally recommended, both on the grounds of computational efficiency and possible matrix conditioning problems. Formally, it requires the inversion of $n \times n$ system of equations, followed by inversion of $m \times m$ system: it is the sparsity of the Hessian matrix that makes the approach effective here. If numerical problems were encountered it would be possible to improve the conditioning of equation (72) by introducing an orthonormal basis for the range space of the sensitivity matrix as discussed by Gill *et al* (1981).

It is also normal to use some form of step length control as an integral part of any Newton method and many methods require some form of move limits. Because the problem formulation using inverse variables becomes unbounded as x_j approaches zero, an arbitrary cut-off requiring $\delta x_j^{(k)} \geq -\frac{1}{2} x_j^{(k)}$ is imposed, limiting the amount any thickness or cross-sectional area may increase during a single iteration. Other than that, the optimisation of membrane structures under static loading appears to work well without recourse to any from of step length control. Application to plate bending structures or, more particularly, to structures subject to dynamic or aeroelastic requirements, tends to introduce more non-linearity into the constraints and step length control designed to address that problem is discussed in section 5.4.3.

The method treats all members of the active set as equality constraints in the first instance, but an active set strategy described below, in section 5.6, is used to delete unwanted constraints.

5.4.2 Relationship with Dual Methods. Equation (70) may be regarded as defining the exact solution of an optimisation model in which a quadratic approximation to the objective function is taken and the constraints are assumed to be linear, thus requiring the minimisation of

$$W = \frac{1}{2}\sum_{j=1}^{n} \delta x_j \frac{2w_j}{x_j^{(k)3}} \delta x_j - \sum_{j=1}^{n} \frac{w_j}{x_j^{(k)2}} \delta x_j + \sum_{j=1}^{n} \frac{w_j}{x_j^{(k)}} \qquad (73\text{ a})$$

subject to

$$g_i + \sum_{j=1}^{m} \frac{\partial g_i}{\partial x_j} \delta x_j = c_i \ . \qquad (73\text{ b})$$

Using equation (71) to eliminate δx_j from the Lagrangian function corresponding to this approximation, gives rise to a function V of the dual variables λ_i, where

$$V(\lambda) = -\frac{1}{2}\sum_{i=1}^{m}\sum_{p=1}^{m}\lambda_p\left[\sum_{j=1}^{n}\frac{\partial g_p}{\partial x_j}\frac{x_j^{(k)3}}{2w_j}\frac{\partial g_i}{\partial x_j}\right]\lambda_i + \sum_{i=1}^{m}\left\{(g_i - c_i) + \sum_{j=1}^{m}\frac{x_j^{(k)}}{2}\frac{\partial g_i}{\partial x_j}\right\}\lambda_i. \qquad (74)$$

This function is the dual objective function to the quadratic programming problem, and the Newton equation (72) defines the maximising point of the dual, and hence the solution to the original problem.

Thus, the Newton method in STARS is based on the used of an optimisation model which makes it equivalent to a sequence of quadratic programs (SQP) and it may also be regarded as the application of a dual method.

5.4.3 Constraint curvature and Step-length control Strictly, if the quadratic convergence associated with Newton's method is to be obtained, the constraints should not be assumed linear in (70a). This desirable convergence is sacrificed in order to simplify the algorithm and to avoid the need to calculate second derivatives of the active constraints. Linearising of the constraints can even cause the algorithm to diverge and so, to prevent such an occurrence, second derivative information has been included in a line search.

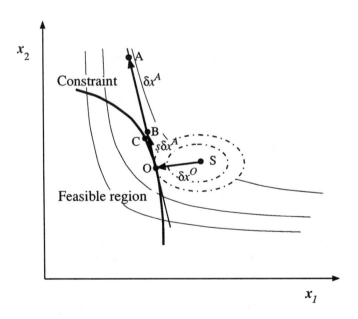

Figure 1 Design space showing the influence of constraint curvature on an optimisation step

STARS: mathematical foundations

If such an *a priori* assumption of constraint linearity is not justified then the algorithm will not converge without some form of step length control. This can be seen from figure 1 where, starting at an infeasible point S, the algorithm will step to the point A if the constraint is approximated by its tangent plane OA, whereas the optimum lies at the point C on the non-linear constraint surface.

To overcome such problems, STARS employs the step length control based on the directional second derivatives of the Lagrangian function, as in Bartholomew (1988). First, the step taken by the Newton algorithm is readily separated into two parts. Setting the right hand side of equation (70a) to zero gives a step δx^O which minimises a quadratic form centred on the current point S (shown dashed in figure 1); that is a least squares restoration step from S to O. Conversely, setting $(c_i - g_i)$ to zero on the right hand side of equation (70b) gives a step δx^A which lies in the null space of $\partial g_i/\partial x_j$; that is a minimisation step from O to A which seeks to minimise the objective function in the constraint tangent space.

This step within the tangent space may be substantially improved by introducing a scaled step $s\delta x^A$ to give B as the next point in the solution sequence. The scale factor itself is determined by the relative curvature of the objective function and the Lagrangian function in the search direction.

Thus,

$$s = \frac{\frac{d^2 f}{ds^2} - \sum_{i=1}^{m} \lambda_i \frac{d^2 g_i}{ds\, dr}}{\frac{d^2 f}{ds^2} + \sum_{i=1}^{m} \lambda_i \frac{d^2 g_i}{ds^2}} \tag{75}$$

where

$$\frac{d}{ds} \equiv \sum_{j=1}^{n} \delta x_j^A \frac{\partial}{\partial x_j}, \qquad \frac{d}{dr} \equiv \sum_{j=1}^{n} \delta x_j^O \frac{\partial}{\partial x_j}.$$

The calculation of directional second derivatives in STARS relies heavily on the use of the approximate analysis techniques outlined in section 3.1.3. These are employed to provide estimates of the first derivatives of response quantities for a perturbed design, and the directional second derivatives follow by differencing.

5.4.4 Generalisations of Newton Method.

The step length control of the last section is one of several approaches to overcoming limitations of the linear-quadratic optimisation model and extending its applicability to a wider class of structural optimisation problem. In that instance the Newton optimisation algorithm is allowed to extend the information inherent in the optimisation model by seeking additional information on second derivatives from the analysis.

An alternative approach is to extend the information represented in the optimisation model. For example, the normal displacements of a structure loaded in bending may be expected to vary as inverse cube of plate thickness. If an approximation of the form

$$u = \sum_{j=1}^{n} (a_j x_j + b_j x_j^3) \tag{76}$$

is assumed for a displacement constraint, the sensitivity will be of the form

$$\frac{\partial u}{\partial x_j} = a_j + 3b_j x_j^2 \tag{77}$$

and the coefficients a_j and b_j are estimated by accumulating contributions to the derivative due to changes in membrane and bending element stiffness matrices separately. The problem remains separable and the diagonal form for the Hessian matrix,

$$H_{jj} = \frac{2w_j}{x_j^2} + 6\lambda b_j x_j \tag{78}$$

enables the Newton algorithm to be applied unaltered, provides care is taken to ensure the terms remain positive. Both this approach and the use of step length controls have been applied with success to a range of beam problems in Bartholomew (1988) and dynamics problems in Bartholomew (1991).

A further generalisation of form for the Optimisation Model which could be made without impairing the efficiency of the method, is to allow a block-diagonal form for the Hessian. Typically this would arise when several dimensions are used to describe a structural component. Currently, work is in hand to enable problems of this kind to be solved using specialised detail design procedures as part of a Multi-level Optimisation.

5.5 Convergence Checking using Duality

While a previous section, 5.4, showed the Newton method to be equivalent to solving a quadratic programming approximation in its dual form, in this section the dual form of the structural optimisation problem is used to establish a lower bound on the optimum weight.

Consider the Lagrangian formulation of the basic problem. The dual problem requires the maximisation of the Lagrangian function (63) with respect to the set of Lagrangian multipliers and primal variables, subject to the satisfaction of equations (64a) as equality constraints. That is

Maximise

$$L(x,\lambda) = \sum_{j=1}^{n} \frac{w_j}{x_j} + \sum_{i=1}^{m} \lambda_i (g_i(x) - c_i) \tag{79 a}$$

subject to

$$\frac{w_j}{x_j^2} = \sum_{i=1}^{m} \lambda_i \frac{\partial g_i(x)}{\partial x_j} \qquad (79\,b)$$

$$\lambda_i \geq 0. \qquad (79\,c)$$

If the original problem is assumed to be convex, then the maximising value for the dual objective function and the minimum value for structural weight are the same. This can be of some value for the development of a solution technique, but the dual formulation also requires the solution of complicated sets of non-linear equations which can be as difficult to solve as the original (primal) problem. However, advantage can taken of the fact that the value of the dual obtained from a non-optimising combination of primal and dual variables provides a lower bound estimate of the optimum structural weight.

To provide a bound at the end of any given iteration, for one of the algorithms outlined above, the design variables are fixed at their current value $x_j\,(k)$, thus rendering both the dual objective function and the dual constraints linear. The resulting problem, discussed further in Bartholomew (1979), is

Maximise

$$V(\lambda) = \sum_{i=1}^{m} \lambda_i g_i(x^{(k)}) \qquad (80\,a)$$

subject to

$$\sum_{i=1}^{m} \lambda_i \frac{\partial g_i(x)}{\partial x_j} \leq \frac{w_j}{x_j^{(k)2}} \qquad (80\,b)$$

$$\lambda_i \geq 0. \qquad (80\,c)$$

This is solved by the application of a standard linear programming algorithm, as documented by Land and Powell (1973), and provides a good lower bound on the optimum value for the structural weight.

For a large and complicated optimisation/automated design system, the dual has an important role to play in monitoring the progress of the various algorithms available within the package. For example, by providing a lower bound estimate for the weight, the iterative process may be terminated when potential design improvements are too small to merit further processing. A graphical display of this "duality gap" assumes a major role in on-line monitoring.

5.6 Active Set Strategy

As we have already indicated, several of the algorithms described above need some form of active set strategy which identifies the constraints forming equalities at a given iteration. There are, however, two distinct phases: an initial active set strategy is required to select constraints on entering the first iteration; and an alternative main strategy is employed on subsequent iterations. The initial active set strategy uses the duality routine outlined above to provide a set of Lagrangian multipliers. The constraints associated with non-zero multipliers are designated as being members of the active set.

The main active set strategy is more involved: at the end of each iteration the violated constraints may be added to the active set whilst those associated with zero or negative Lagrangian multipliers may be deleted. However, in order to avoid "zig-zagging", it is customary to use a more conservative deletion policy. One possibility, for algorithms which develop their own Lagrangian multipliers such as the Newton method of section (5.4), is to delete the constraint having the largest negative multiplier at each iteration. In STARS, an analysis is performed on the set of negative multipliers to choose for deletion the constraint which gives the greatest reduction in the dual objective function (74).

An additional strategy, aimed at stabilising the active set, is to place a limitation on the number of stress constraints added at each iteration. A stress constraint is only included in the active set if it is violated and corresponds to the most highly stressed member associated with a design variable. It is assumed that the inclusion of other associated constraints would lead to undesirable linear dependencies within the constraint set.

6. EXAMPLES

6.1 Cantilever Beam

The first example is a model of a simple cantilever beam. On the face of it, this example would appear to be so simple that the use of sophisticated optimisation tools would be

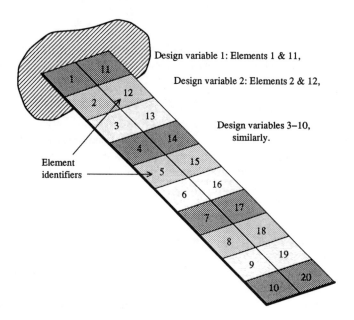

Figure 2 Cantilever Beam showing design variables

unwarranted. However, the model is constructed so as to demand the simultaneous satisfaction of static, natural frequency and forced response constraints. These constraints make the problem difficult to solve manually. Figure 2 is a plot of the finite element mesh with the design variables linking indicated, to show how they have been chosen as bands along the beam's length. The optimisation results, given in figure 3, show that convergence is obtained in only a few iterations, with all constraints being satisfied.

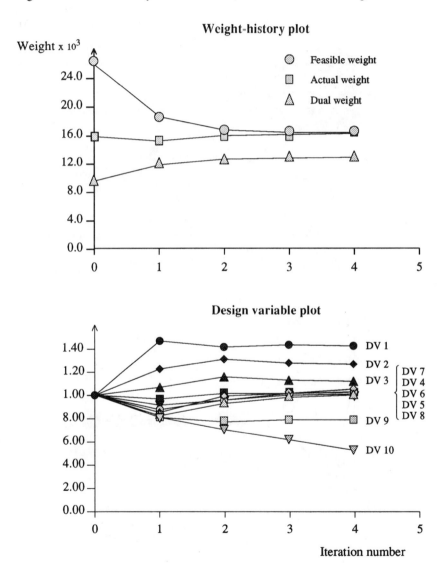

Figure 3 Iteration History for Cantilever Beam

6.2 Helicopter

With this example of a model of the Westland's LYNX helicopter, the objective is to use structural optimisation techniques to assist in the task of passive vibration reduction at key points in the structural model.

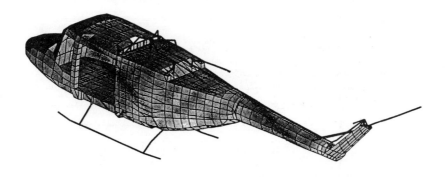

Figure 4   Finite element mesh of Helicopter

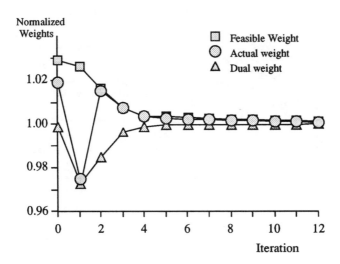

Figure 5   Iteration History for Helicopter

STARS: mathematical foundations 281

Figure 4 shows an isometric view of the model which had approximately 9000 degrees of freedom. This free-free model is subjected to a forced response loading associated with a particular forward speed of the helicopter and the design variables are defined in bands running from the tail through to the nose cone. Each band is then sub-divided into two regions, one above the centre line and the other below. Beams and skin panels are allowed to scale independently. The principal constraints are associated with accelerations at the key points, but more stable convergence is achieved if natural frequency constraints are also imposed on modes above the forcing frequency. A 10% reduction in vibration is achieved, with the weight optimisation history shown in Figure 5.

6.3 Mounting Bracket

This example consists of a mounting bracket which is subject to stringent stiffness and natural frequency constraints. It consists of three lips which are fully built in at their edges remote from the bracket, as illustrated in Figure 6, and a stiffened plate which is offset from the mounting points by side walls modelled using plate bending elements. Two static load cases are applied to the initial structure which had a fundamental frequency of 7.5Hz. Constraints are applied to reduce the static displacement to under 0.5mm at all nodes and to increase the fundamental frequency to above 12.5Hz.

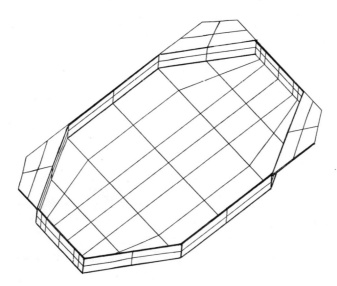

Figure 6 Finite Element model of the Mounting Bracket

The results of the analysis, shown in Figure 7, indicates a converged solution after eight iterations. The natural frequency constraint is rapidly satisfied and then the subsequent iterations are required to resolve the stiffness requirements. Most material is added to the lip elements adjacent to the supports but the base elements and the side walls are also thickened to enable the overall stiffness of the bracket to be increased and so enable the design constraints to be met.

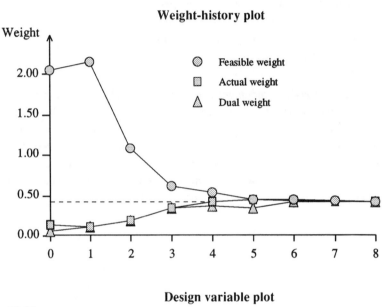

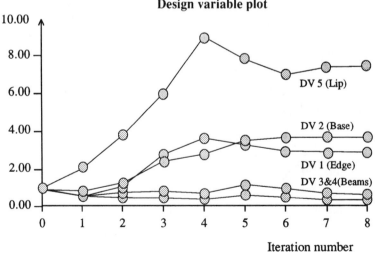

Figure 7 Iteration History for the Mounting Bracket

REFERENCES

Allman D J (1976), "A simple cubic displacement element for plate bending", *Int J Numerical Methods in Engineering*, Vol 10, pp262-281.

Atrek *et al* (Eds.) (1984), *New directions in optimum structural design*, John Wiley.

Auld A B K, Knott M J, Lock W E and Sullivan M C W (1986), STARS Theoretical Manual, Aeroelastic Supplement, British Aerospace, BAe-KSD-R-GEN-1304

Bartholomew P (1979), "A dual bound used for monitoring structural optimisation programs", *Engineering Optimisation*, Vol 4, pp45-50, (1979).

Bartholomew P and Pitcher N (1984), "Optimization of Structures with repeated normal-mode frequencies", *Engineering Optimization*, Vol 7, pp195-208.

Bartholomew P (1988), "Computer Aided Structural Design", pp 285-309, *Simulation and Optimization of Large Systems*, Ed A J Osiadacz, IMA Conference Series.

Bartholomew P (1990), "A New Aproach to the Optimistion of Structures subject to Frequency Constraints", *31st AIAA SDM Conference, CP902*, Long Beach CA

Bartholomew P (1991), "Calculation of Eigenvectors of Modified Systems for use in Structural Optimisation", *Int Forum Aeroelasticity and Structural Dynamics*, Aachen.

Corr R B and Jennings A (1976), "A simultaneous iteration algorithm for symmetric eigenvalue problems", *Int J Numerical Methods in Engineering*, Vol 10, pp 647-663.

Eschenauer H A (1989), "The three columns for treating problems in optimum structural design", *Optimisation: Methods and Applications, Possibilities and Limitations*, Ed Bergman H W, DLR Bonn, Springer Verlag.

Gill P, Murray W and Wright M (1981), *Practical Optimization*, pp183, Academic Press Ltd, London.

Land A and Powell S (1973), *Fortran codes for mathematical programming*, John Wiley.

Morris A J (Ed.) (1982), *Foundations of Structural optimisation: A Unified Approach*, John Wiley.

Schmitt L and Pope G G (Eds) (1973), AGARD *Second symposium on Structural Optimisation*, Milan.

Zienkiewicz O C (1977), *The finite element method* (Third edition), McGraw-Hill.

Authors' address:

P Bartholomew
Defence Research Agency
Materials and Structures Dept
Farnborough
Hants GU14 7TD
Great Britain

Sarah Vinson
EDS-Scicon Ltd
Wavendon tower
Wavendon
Milton Keynes MK17 8LX
Great Britain

Titles previously published in the series

INTERNATIONAL SERIES OF NUMERICAL MATHEMATICS
BIRKHÄUSER VERLAG

ISNM 93   R.E. Bank, R. Bulirsch, K. Merten (Eds.): Mathematical Modelling and Simulation of Electric Circuits and Semiconductor Devices, 1990 (3-7643-2439-2)

ISNM 94   W. Haussmann, K. Jetter (Eds.): Multivariate Approximation and Interpolation, 1990 (3-7643-2450-3)

ISNM 95   K.-H. Hoffmann, J. Sprekels (Eds.): Free Boundary Value Problems, 1990 (3-7643-2474-0)

ISNM 96   J. Albrecht, L. Collatz, P. Hagedorn, W. Velte (Eds.): Numerical Treatment of Eigenvalue Problems, Vol. 5, 1991 (3-7643-2575-5)

ISNM 97   R.U. Seydel, F.W. Schneider, T.G. Küpper, H. Troger (Eds.): Bifurcation and Chaos: Analysis, Algorithms, Applications, 1991 (3-7643-2593-3)

ISNM 98   W. Hackbusch, U. Trottenberg (Eds.): Multigrid Methods III, 1991 (3-7643-2632-8)

ISNM 99   P. Neittaanmäki (Ed.): Numerical Methods for Free Boundary Problems, 1991 (3-7643-2641-7)

ISNM 100 W. Desch, F. Kappel, K. Kunisch (Eds.): Estimation and Control of Distributed Parameter Systems, 1991 (3-7643-2676-X)

ISNM 101 G. Del Piero, F. Maceri (Eds.): Unilateral Problems in Structural Analysis IV, 1991 (3-7643-2487-2)

ISNM 102 U. Hornung, P. Kotelenez, G. Papanicolaou (Eds.): Random Partial Differential Equations, 1991 (3-7643-2688-3)

ISNM 103 W. Walter (Ed.): General Inequalities 6, 1992 (3-7643-2737-5)

ISNM 104 E. Allgower, K. Böhmer, M. Golubitsky (Eds.): Bifurcation and Symmetry, 1992 (3-7643-2739-1)

ISNM 105 D. Braess, L.L. Schumaker (Eds.): Numerical Methods in Approximation Theory, Vol. 9, 1992 (3-7643-2746-4)

ISNM 106 S.N. Antontsev, K.-H. Hoffmann, A.M. Khludnev (Eds.): Free Boundary Problems in Continuum Mechanics, 1992 (3-7643-2784-7)

ISNM 107 V. Barbu, F.J. Bonnans, D. Tiba (Eds.): Optimization, Optimal Control and Partial Differential Equations, 1992 (3-7643-2788-X)

ISNM 108 H. Antes, P.D. Panagiotopoulos: The Boundary Integral Approach to Static and Dynamic Contact Problems. Equality and Inequality Methods, 1992 (3-7643-2592-5)

ISNM 109 A.G. Kuz'min: Non-Classical Equations of Mixed Type and their Applications in Gas Dynamics, 1992 (3-7643-2573-9)

ISNM

A series with a long-standing reputation

Since its foundation in 1963 more than 100 volumes have been published by Birkhäuser Verlag in the **International Series of Numerical Mathematics**.

John Todd's *Introduction to the Constructive Theory of Functions*, published as Volume 1, was a remarkable start. Proceedings volumes and further monographs such as Fenyö/Frey, *Moderne mathematische Methoden in der Technik*, Ghizzetti/Ossicini, *Quadrature Formulae*, Todd, *Basic Numerical Mathematics* (two volumes) and Heinrich, *Finite Difference Methods on Irregular Networks* followed, always presenting the state of the art in exposition and research.

Originally the Editorial Board consisted of Ch. Blanc, A. Ghizzetti, A. Ostrowski, J. Todd, H. Unger, A. van Wijngaarden. Despite a number of changes, it has shown long years of continuity; Prof. Ostrowski and Prof. Henrici, for instance, had been members of the Board all their life.

At present the series is being edited by
Karl-Heinz Hoffmann, München, **Hans D. Mittelmann**, Tempe, **John Todd**, Pasadena.

As in the past, we do not intend to restrict the series a priori to certain subjects. The series is open to all aspects of numerical mathematics. At the same time, we wish to include practical applications in science and engineering, with emphasis on mathematical content.

Some of the topics of particular interest to the series are:
Free boundary value problems for differential equations, phase transitions, problems of optimal control and optimization, other nonlinear phenomena in analysis;
nonlinear partial differential equations, efficient solution methods, bifurcation problems; approximation theory.

If possible, the topic of each volume should be discussed from three different angles, namely those of Mathematical Modelling, Mathematical Analysis, Numerical Case Studies.

The editors particularly welcome research monographs; furthermore, the series is to contain advanced graduate texts, dealing with areas of current research interest, as well as selected and carefully refereed proceedings of major conferences or workshops sponsored by various research centers. Historical material in these areas would also be considered.

We encourage preparation of manuscripts in LaTeX or AMSTeX for delivery in camera-ready copy which enables a rapid publication, or in electronic form for interfacing with laser printers or typesetters.